DevSecOps

A leader's guide to producing secure
software without compromising flow,
feedback and continuous improvement

GLENN WILSON

R^ethink

First published in Great Britain in 2020
by Rethink Press (www.rethinkpress.com)

Contents

This book is dedicated to Caz, Abs and Emily

Foreword

I have observed the dramatic changes in the software development field through a security lens during the 2010s. The pace at which the business demands rapid releases of new features to capture and accelerate business growth has become the norm.

This explosion is obvious when you look at the extraordinary rise of the likes of Etsy, Amazon, Facebook, Netflix, Taobao and WeChat. Each platform caters for the growth and spike in traffic, and the constant releases of new capabilities and updates to their web and mobile interfaces.

This fast pace of multiple releases (per day in most cases) has led the industry to accept that a security breach is going to happen to any business and so organisations must be prepared to handle such scenarios. As a counter to this inevitable event, organisations are now driving their delivery teams even harder to not just deliver on time but to provide a quality that exceeds the standard of their competitors – and this includes security. Reputation damage and staying out of the press due to a security breach is now a top agenda item for the executives. Security has finally become a first-class citizen.

Being in the right place at the right time (through my commercial engagements) has allowed me to experience the evolution first-hand. Watching how the software development practices rapidly adapt to the business needs through the creation of new vocabulary and technology, turning software engineering into a sexy field with more and more people aspiring to join, has been an exciting time in my career. During the same period the security industry has achieved different successes; for example, the role of a security representative – Chief Information Security Officer (CISO) – on the board of companies, the acceptance of different shades of penetration testing (red, blue, purple and bug bounties) as an important assurance stage, and the formation of different teams covering the various security domains and disciplines required to serve the business.

In this book Glenn highlights the core principles in modern software development and security. He draws upon both his experience as a security practitioner and the wisdom of other leading figures in the area to merge the two practices, explaining the need and demand for what we now call DevSecOps.

Glenn proposes a three-layer approach for tackling DevSecOps.

Staying up to date with the current practices and emerging technology, and learning from other people's success and mistakes, is a never-ending journey. Each individual's experience and skillset are different and so, in the first DevSecOps layer, educational programmes must be devised and adaptable to allow each resource to understand the core principles of DevSecOps. This is to ensure the security fundamentals are well understood by, and ingrained into, the delivery team. Glenn discusses different learning methods to offer teams, ranging from gamifying online training to running tournaments and establishing security champion roles to allow people to put theory into practice – reinforcing the lessons taught. He notes that teaching materials are also available outside of the company. The 'teaching programme' adopted for each individual should reflect how they learn and their willingness to expand their security knowledge.

The second layer focuses on the design aspect of the solution. Glenn covers good and secure coding practices, such as threat modelling and peer programming. In addition, he shares respected industry references from the OWASP and SANS establishments. The details from these will complement the lessons learned, prescribed from the training programme, and hopefully start to bake security into the core practices of the delivery team. The adopted software architecture for the solution must also be supportive of the modern development practices, and so the world of containers and microservices technology are also discussed. The reliability of the delivery pipeline and its security becomes more important in the world of DevSecOps.

This layer concludes by highlighting good practices for securing the delivery pipeline.

Completing the trio, the third layer discusses the different security testing techniques which can be deployed as part of your DevSecOps practice. Automating the deployment and testing is key, but, in reality, not all testing can be automated due to certain industry regulations. Most teams will challenge the need for traditional penetration testing, but the use and purpose of this activity is not dismissed from Glenn's three-layer framework.

To bring everything together, Glenn wraps up the book by structuring a programme of activities for teams to explore and adopt to start their journey into the world of DevSecOps.

This book will no doubt become a mandatory reference in the DevSecOps culture – I hope you will enjoy it as much as I have.

Michael Man
DevSecOps London Gathering
September 2020

Introduction

Designing applications, whether they are simple APIs (application programming interfaces), microservices or large monoliths, is a complex activity involving resources from many disciplines. In traditional software delivery practices involving a slow linear lifecycle (such as 'waterfall'), a product starts as an idea to meet a perceived requirement of the end user. This idea is handed over to a team of architects who design the application according to a set of criteria that meet the business objectives. The architecture determines the key features of the software and infrastructure, as well as non-functional requirements such as performance. When the architecture is complete, software engineers write the code that makes these pre-designed features, while operations engineers build the underlying infrastructure and services needed to host the applications. Finally, teams of testers validate the application against the functional and non-functional requirements and provide the quality assurances that meet user expectations.

These traditional methodologies have given ground to newer ways of working that merge the roles of product owners, architects, developers, testers and representatives of end-users into integrated product teams providing faster delivery and greater agility and responsiveness to changes in requirements. Furthermore, cloud hosted applications have led to the inclusion of operations resources within multidisciplined teams to write infrastructure as code, which enables products to be developed, tested and deployed as complete packages. These teams and their methodologies have come to be known as *DevOps* (from the combination of *dev*elopment and *op*erations). The necessity of delivering new products and features to customers as quickly as possible has driven the journey from traditional waterfall to DevOps. Organisations that struggle to adapt to new ways of working have learned that the lengthy waterfall processes are unable to compete against the more agile and responsive organisations.

A major challenge facing the industry is understanding how security practices fit into the newer models. In traditional ways of working, project artefacts are handed from one team to the next when specific milestones have been met. Security integrates into this process with several distinct steps, each requiring the project to validate security controls and document open risks. Thus, the conceptual design is assessed through threat modelling activities for potential security risks. Once the architecture has been fully assessed and open risks have been fully documented, the project advances to the development stage. Developers implement the relevant security controls based on the initial design document and pass on the application to a team of testers.

The testing phase may include a dedicated security tester, although this activity normally takes place after functional defects have been identified and fixed and the application is ready for deployment to a live system. Security testing may even be carried out by an external team. Ultimately, the output is a report of vulnerabilities normally classified by criticality. The project delivery manager is cornered into making a decision on whether to delay the project while remediation takes place or to accept the risks and deliver the project without remediation.

Finally, the security operations centre (SOC) continues to monitor the application for security incidents using the project's artefacts, such as the original design documents, vulnerability reports and user guides, as reference points for the application. Although this methodology has its faults, it is a familiar story for many delivery managers working on software projects. Moreover, security departments have invested in this approach by creating teams of specialists covering a number of security disciplines needed to support project delivery. These specialists can include threat modellers, security testers, firewall administrators, key infrastructure and certificate management teams, and identity and access control teams.

The Agile framework (allowing teams to deliver working software in short delivery cycles) and DevOps offer the opportunity to integrate security as part of the ongoing feedback loop. However, in order for this to happen, the traditional central cybersecurity roles need to change in order to continue supporting the delivery of secure software solutions. As Eliza-May Austin, of th4ts3cur1ty.company and co-founder of the Ladies of London Hacking Society, points out:

'Developers have done phenomenal work in the DevOps space, whereas Security has not kept up with them. It's not feasible for DevOps to slow down, so security needs to step up. This is easier said than done since it requires a cultural shift in the way the software delivery teams, and security teams work together.'

The lack of security engineers with experience in Agile and DevOps within an organisation means that they are not being integrated into the DevOps teams. Furthermore, DevOps teams fall short in the level of knowledge required to integrate security into their ways of working. The result: security teams are treated like external resources; and DevOps and Agile teams are forced to reach out to security teams to configure firewalls, manage certificates and keys, and put in place access controls to support the running of the products. Compounding matters, the delivery team is still likely to hand the product over to a SOC team for ongoing support, creating even more distance between security and DevOps, which ultimately creates a slow feedback loop. The constraints forced upon DevOps teams by security departments provide the catalyst for managers to look for quick wins, often at the expense of security.

A potential solution is to integrate security engineers into the DevOps teams, bolstering the knowledge and skills within the team to address this challenge. Unfortunately, that solution does not scale. Typically, the ratio of security engineers to software engineers is very small; estimates range from one security engineer per 100 developers to one per 400. Overworked security engineers struggle to meet the ever-increasing demands of the software delivery teams who are delivering products and product features at an accelerating pace. Therefore, we cannot simply add a security engineer to a DevOps team and call it *'DevSecOps'*. A different approach is required – one that is scalable and effective.

Some may argue that if DevOps is done correctly, security is implicitly applied. There are two problems with this concept. Firstly, DevOps has different meanings to different people. Some engineers focus on the automation of operations to define DevOps, while others focus on cloud services to define DevOps. Secondly, security is more complex than just doing DevOps right. There are many factors influencing security best practices which need to be explicitly defined. The rest of this book proposes a solution to this problem: I introduce a three-layered approach to embed a culture of security and security practices within the organisation, putting security at the forefront of DevOps to create Dev*Sec*Ops.

I define the *three layers* of DevSecOps that help decision makers integrate security into DevOps ways of working. DevSecOps increases the safety of the products your organisation develops while continuing to deliver new products and features that give your company a competitive edge.

ONE

DevOps Explained

'DevOps' – the word – is simply the merging of the first letters of two distinct functions within IT: the *Dev*elopers, who are responsible for writing software, and the *op*erations staff, who are responsible for maintaining the infrastructure on which the software is developed and deployed. Of course, 'DevOps' – the function – is far more complex than a merging of two roles into a single team. To understand what DevOps is, you need to delve into its history. Traditional software development is based on the waterfall methodology, framed within a project that starts with ideation and ends with the delivery of a working application. There are many problems with this delivery approach, the main one being the lack of flexibility. Often, the requirements change over the course of the project, leading to increased scope, missed milestones and higher costs. To keep projects on track, it is not uncommon to see a lack of quality in the final project deliverable. The constraints associated with scope, time and cost, and their effect on quality, are represented below.

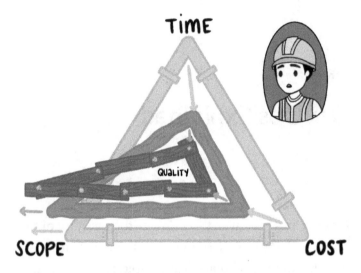

Constraint triangle: the effect of increased scope and reduced cost and time on quality

Furthermore, customer requirements are based on vague assumptions made by various stakeholders during the project lifecycle, which are only validated once the project has ended and the users are finally able to interact with the project deliverable. By this stage, the project team has already been disbanded and handed over (with supporting documentation) to a support team, which restricts product changes to fixing bugs or providing workarounds for users to negotiate poor functionality. These changes often bloat the software with awkward fixes, making the code more difficult to support, ultimately increasing technical debt.

The industry's solution to the problem of slow project delivery was two-fold. The first change was taken from the car manufacturing industry that streamlines its production lines by using lean practices. Lean manufacturing uses small work packages, just-in-time supply chain management and automation of repeatable processes to reduce waste, such as extended lead times and ineffective work. Keeping the processes lean and product-focused promoted team structures and workflows based on the end-to-end tasks required to deliver value to the customer for a specific product. The second part of the solution came from a group of software engineering experts who produced a manifesto with a set of software developing rules that focuses on the values of individuals and interactions, working software, customer collaboration and responding to change. This *Agile Manifesto* defines a set of twelve principles that promote quality, continuous delivery (CD), regular feedback loops and collaboration between individuals and teams. Various Agile methodologies evolved from this movement, including test automation and test-driven development (TDD), pair programming and continuous integration (CI).

As these ideas developed further, and software development became more streamlined and product-focused, bottlenecks shifted from product development to infrastructure support. Once software was ready to be delivered to an appropriate environment, such as a test environment or pre-production environment, or to a full production environment, the packages were handed over to a team that managed deployments. To remove this bottleneck, infrastructure was automated using code and the operations tasks needed to build the infrastructure were integrated into the Agile development processes. CI extended to CD to create a pipeline that built, tested and delivered the whole package: the application, the services in which the application runs and the infrastructure on which they are all hosted. The integration of development and operations into lean and agile practices is known as *DevOps*.

DevOps teams are made up of a number of other disciplines, including testers, architects and product owners as well as the developers and operations staff. Each DevOps team is able to work as a single unit with minimal dependencies on other teams. There may be some interaction between DevOps teams, but ideally each DevOps unit works in isolation. There is no concept of handing off to another department to perform a function. This self-sufficient team can design a feature, write and test code, generate the environment on which the software runs and deploy the whole package into a production environment, all while looking for ways to continuously improve this process.

In the following sections, we will briefly explore the three ways and the five ideals that frame the DevOps movement.

The three ways

In the seminal work *The DevOps Handbook*, co-authors Gene Kim, Jez Humble, Patrick Debois and John Willis describe three principles underpinning DevOps. They built these principles for the software engineering industry by examining successful production lines within the manufacturing industry and evolving best practices for developing software. The first of their principles is the left-to-right process flow. In this principle, the focus is on delivering features in low-risk releases by incorporating automated testing and CI into the deployment pipeline. The second principle is based on using right-to-left feedback mechanisms that allow engineers to anticipate and resolve problems rather than waiting for problems to occur in a production environment. The third principle provides an environment for continuous learning and experimentation, allowing engineers to continuously improve development and operations as an embedded part of the process. These three principles (or ways) are the foundations of DevOps.

Three ways of DevOps: left-to-right process flow, right-to-left feedback and continuous improvement

The five ideals

Following the release of *The DevOps Handbook*, Gene Kim's *The Unicorn Project* extends the principles of DevOps into five ideals that collectively define the core values of DevOps. The first of these ideals is *locality and simplicity*. In order for a team to independently build, test and deploy value to customers, it needs to avoid having dependencies on a large number of other teams, people and processes. Each DevOps team can make its own decisions without needing a raft of approvals from others, it promotes the decoupling of components to simplify development and testing, and it recommends making data available in real time to those who need it to complete their tasks efficiently.

The second ideal states that teams must have *focus, flow and joy*, meaning they must be free from constraints that hinder their ability to complete their tasks. Individuals who have to work on multiple activities at the same time or have multiple disruptions while working on an activity are less likely to work to a high standard. If teams are able to focus on individual actions without interruptions, they gain a sense of joy from being able to complete their work. This helps the team deliver value to customers.

Improvement of daily work is the third ideal; it focuses on the reduction of technical debt, including security weaknesses. Technical debt, if not addressed, will grow to such an extent that most or all daily work will need to work around it to deliver features, fix defects and mitigate risks. Therefore, a significant proportion of the team's daily work must involve investing in developer productivity and reducing technical debt. Within *The DevOps Handbook*, Kim identifies four types of work: business related (such as creating new features), IT, infrastructure improvement and unplanned work. Unplanned work becomes a distraction from delivering value and is symptomatic of large amounts of technical debt.

DevOps teams should not be fearful of being open and honest, which is the essence of the fourth ideal, *psychological safety*. Rather than evolving a 'blame, name, shame' culture, individuals must be able to speak up without fear of repercussions. It is important to reinforce a culture of openness within a DevOps team so that if a problem is identified, it is exposed and fixed.

The final ideal is *customer focus*. DevOps teams have two types of customers: external (the paying customer who benefits from the features that are delivered to them) and internal (individuals or teams who receive outputs from those immediately preceding them in the value stream). By focusing on the immediate consumer of your output, there is greater opportunity for instant feedback, leading to greater customer experiences.

Conclusion

DevOps was born from the need to accelerate product development to gain an advantage over competitors. I have seen the word 'DevOps' used arbitrarily to describe Agile methodologies used during development and justified by the use of certain processes, such as CI, automated testing and containerisation. However, I have also witnessed scenarios in which DevOps is framed within a project, teams stick rigidly to predefined processes, and individuals within DevOps teams act as heroes to keep the project on track. In this book, I endeavour to remain faithful to the true meaning of DevOps as outlined in this chapter, while ensuring that the principles I introduce in the rest of the book can be applied to less mature DevOps practices as well as more mature DevOps organisations.

TWO

Security Explained

Hardly a week goes by without some form of security incident making news headlines, from large financial institutions announcing a data leak involving thousands, if not millions, of stolen customer credit card numbers, to government offices unable to serve their citizens due to an ongoing ransomware attack. Unfortunately, these happen far too frequently and affect many people, including your customers and employees as well as specific groups with the general population. In most cases, the attackers were able to exploit weaknesses in the organisation's public-facing network due to vulnerabilities in the underlying software or configuration issues within the environment on which the application is hosted.

Implementing security hygiene throughout your organisation has become more critical than ever and security is a topic that is as broad as it is deep. There are many ways to exploit an organisation through myriad weaknesses: theft of data stored on laptops, removable media or even paper documents left in poorly secured locations or in public places; passwords exposed through social engineering or phishing emails; or customer data stolen from poorly secured data stores or via weaknesses in web-based applications. Security keeps your valuable assets safe and keeps intruders with criminal intentions away from your critical systems. In software engineering, keeping your products and services secure has become increasingly complicated as more devices connect to the internet and users have instant access to your online offerings from anywhere in the world. This opens up greater opportunities for attackers to exploit more devices and enables more users to steal valuable information. In this chapter, I explain the types of attacks that may affect your organisation, the types of adversaries that may attack your organisation and why security is important in software engineering.

Types of attacks

Attacks come in different guises, including phishing attacks normally used to obtain user credentials, an advanced persistent threat (APT) to harvest information from a system over a period of time, or a denial of service (DoS) attack. There are many attack scenarios, each with a specific goal. You may not always be the intended target: some organisations suffer from collateral damage as a result of a partner or supplier coming under attack. However, many attacks *are* targeted. In these cases, a determined attacker will spend many days, weeks or months finding a way to hack a specific organisation. In other scenarios, the attackers scatter their attacks across the internet to compromise as many machines as possible. For example, viruses are able to infect a system, replicate and spread to other systems. Their motives for creating a 'botnet' (a network of compromised systems) are usually to use the accumulative computing resources to harvest Bitcoins, send large quantities of unsolicited emails or enable a distributed denial of service (DDoS) attack.

The fight against those wishing to do harm through hacking is drawn across multiple constantly changing battle lines. Engineering teams need to be armed with the tools and skills to maintain the security of their products and services at all times. There are three information security principles which should form the foundation of your organisation's security policy. These are *confidentiality*, *integrity* and *availability*, commonly referred to as the CIA triad. Some would add a fourth element: *trust*, although the motives for exploiting trust are directly linked to the other three. Let's delve a little deeper into the CIA triad.

Attacks on confidentiality

In the digital world, our personal data lives in many different places. Our banks hold our financial data, which we can access via online services such as mobile banking, social media companies have information on our personal and professional relationships, and online retailers have data on our purchasing history and often store payment details for the goods we buy. As consumers, we empower and trust these organisations to protect our data and prevent them from falling into the wrong hands – to keep our data *confidential* – yet many of us are victims of a data breach, often unknowingly. Troy Hunt's Have I Been Pwned website reveals an enormous number of email addresses and passwords that have been stolen and published on the dark web, which is a part of the internet hidden from normal web browsing tools and search engines. We think our usernames and passwords are confidential, but many people, including your customers and potentially your employees, have their credentials published by hackers. We also hear about stolen credit card details that have been harvested during a breach or learn that confidential information about our private lives has been leaked by a website we thought we could trust. One popular social media company was affected by a breach in which millions of records were harvested for commercial and political reasons. Attackers can amass large quantities of money either by selling the information or by exploiting the details to steal money from an unsuspecting user.

The responsibility for denying attackers access to data managed by the applications and services your teams develop sits within the engineering team. Those developing the products and services that collect and store sensitive data must keep the data confidential at all times. Accountability for protecting the data's confidentiality sits firmly within the organisation's leadership team, who provide the funds and governance to implement effective controls to protect the data. I believe that both engineers, who write software, configure environments and administer databases, and the organisation's leadership team have a moral duty to protect users' data. It is the responsibility of the production teams to use the most effective controls available to them, while leaders are accountable for ensuring the organisation has the capability to protect users' data.

Encryption prevents attackers from seeing the data in a clear text format. Encryption policies must be supported by an effective and robust key management policy to protect the keys used to encrypt – and, more importantly, to decrypt – the data. Attention should also be given to the algorithms used to encrypt and decrypt data, which must be strong enough to prevent hackers from cracking them. I would suggest that your organisation should never try to create and use its own encryption algorithms. Standard, publicly available algorithms developed through academic research are more secure than any that your engineers can create.

Be aware that not all data needs to be protected. Indeed, there is a trade-off between performance and the processor-hungry encryption and decryption algorithms. Classifying the data based on its sensitivity allows engineers to focus on the more highly restricted data, while public data can remain unencrypted. Encryption and key management are huge topics on which many resources have been written, so this book will not go into specific details. However, it is worth promoting a basic knowledge of encryption among your engineers: they should know the difference between public and private keys, hashing and encoding (which is not a form of encryption although is often mistaken as such). They will also benefit from learning how keys should be transferred between users in order that data can be safely exchanged between systems.

Confidentiality does not relate solely to user data (such as customers, employees and suppliers). Many artefacts within an organisation should remain confidential. These include system passwords which are often used to gain privileged access to a production system. Privileged Access Management (PAM) systems are designed to provide short-lived or one-time passwords to relevant employees under strict criteria, such as investigating an incident or applying an emergency patch. PAM systems are a control to prevent attackers from gaining privileged access to the production system. Your engineers should avoid creating privileged credentials and sharing them with each other or on unprotected internal systems including email, intranet pages and file systems.

Other *confidential* artefacts that need protecting include source code, configuration files and libraries of the products and services that are produced by the engineering teams. These are targets which attackers exploit by gaining access to these artefacts to insert malicious code, backdoors or Trojans into your organisation's products and services. To avoid such breaches, it is paramount that exclusive access to these resources is restricted to the teams and individuals that need access. Role-Based Access Controls (RBAC) define the roles associated with particular functions, such as developers, database administrators and testers. Individuals within these functions are assigned to the relevant roles within the RBAC data store and policies applied to the roles defining the resources to which they have access. Thus, developers may have write-access to a source code repository for the product they are working on, while testers will have read-only access to the same repository. The use of public repositories can provide greater risk to your organisation if not configured to limit access to specific individuals and teams within your organisation.

Maintaining confidentiality of a product or service, and the resources and artefacts used to develop and build it, is the first of the three main controls against attackers.

Attacks on integrity

The second type of attack is an attack on *integrity*. Maintaining the integrity of a system involves ensuring the data behaves in a controlled and predictable way.

This means that, when data flows through an application, from a frontend user interface (UI) to the backend system, or is stored in a database, the data can only be altered in a controlled and expected way. Better still, the original data should remain completely unaltered, and only copies of the data should be manipulated to meet business requirements. Unfortunately, many systems allow changes to data which opens the door to potential attacks on the integrity of the data. For example, an attacker may be able to intercept a transaction within an order and alter the delivery address without the user knowing. This means the buyer pays for an item that is sent to an address chosen by the attacker. The retailer will likely have to either reimburse the buyer or dispatch another item to the correct address at no extra cost, which directly impacts their profit.

The attacker could also intercept a file being uploaded and replace it with another file that contains malicious content. If the unsuspecting victim downloads and opens the file – or, if it is an application, runs the file – with the malicious code, it could cause considerable or irreparable damage to the user's system. Such an attack can also cause reputational damage to the host of the original file.

These attacks exploit weaknesses affecting the integrity of data within a system to the advantage of the attacker. There are a number of controls to mitigate against this type of attack; for example, creating immutable objects that are instantiated with the correct state from classes with no accessors reduces the risk of the attacker intercepting an order to change the address. The object cannot be changed; it can only be read. Extending this further, if your DevOps engineers develop servers or services and deploy them as containers, they should follow best practices that recommend containers remain immutable for their lifetime. Remote access to these containers should be avoided.

DevOps promotes the use of automation to remove the need for manual changes to systems. All changes that are made to an application should be delivered through the CI/CD pipeline. Not only should these changes be automated, any changes in production should be audited. Keeping a record of what the change is, who made the change and for what business outcome the change was made can help identify unauthorised changes to the system.

Another control to maintain the integrity of the system involves generating a hash value from a file that has been downloaded – normally referred to as a *checksum*. This checksum is compared to an expected hash value already generated from the file. A hash is an encoded value created from the file that needs to be protected; it is a one-way encryption technique using a hash algorithm such as Secure Hash Algorithm (SHA). If the hash value of the original data is identical to the hash value of the data being accessed, the end user is assured that the file being downloaded has not been tampered with. If the hash values are different, the file being accessed is not the same as the original expected file.

Maintaining data integrity is an important function

of the DevOps engineering team. Data is at risk from being manipulated and altered whether it's in motion, such as flowing through the application or the deployment pipeline, or at rest when stored in an application data store or a source code repository. All members of the team must identify ways to protect data from malicious or accidental manipulation.

Attacks on availability

Today's consumers demand *availability*. They have an expectation that online services are available on demand, at any time of day and on any day of the year. Whether they are watching a film or listening to music through an online streaming service such as Netflix or Spotify, accessing their online bank account, or buying products from an online store, consumers expect a seamless user experience 24/7. Even when users are browsing sites for information, they expect the information to be immediately available. Consequently, if a user tries to access your organisation's services and they are not responding quickly enough, or not responding at all, they will likely go elsewhere. If your organisation generates revenue through your websites, either directly through online payments or through indirect revenue streams such as advertising, not being able to deliver content in a timely manner can damage your company financially.

You must guarantee your organisation's online services are available as close to 100% of the time as possible. As discussed above, the DevOps model is designed for minimal disruption to customers; when DevOps is fully functioning within an organisation, releases are automated through the deployment pipeline, which undergoes constant review and improvement, while experimentation improves the commercial offering to the consumer. However, a significant attack on the service, affecting its availability to consumers, could negate the inherent benefits of DevOps in minimising downtime. For example, an attacker could disrupt the service by orchestrating a DDoS attack on it. DDoS attacks are designed to overwhelm a service by sending it so many dummy requests that it is no longer able to process legitimate requests efficiently, including traffic from your customers. Eventually, it can cause the service to stop entirely. Attackers often have huge botnets at their disposal to launch DDoS attacks. These botnets are collections of internet-connected devices that have been infected with malware designed to send requests to a particular IP (internet protocol) address on command. For a company targeted by these types of attacks, the outcome can be devastating.

Unfortunately, many botnets offering DDoS services are available on the dark web for a small fee, but there are likewise numerous controls that protect your business from DDoS attacks. Some are built into firewalls while others involve caching data in a separate server. Both types of defences control traffic flowing to the internal network. Developers and operations share responsibility for defending against DDoS attacks by implementing controls across the Open Systems Interconnection (OSI) layers, including the application, session, transport and network layers.

Another kind of attack on availability, and one that has become more common in recent years, is ransomware attacks. These normally extort money from a victim by taking services offline. The only realistic way for an affected organisation to bring the services back online is to pay a sum of money, usually in Bitcoins, to the perpetrator. The most common method of disrupting services is to encrypt all files within the target environment, including application files, configuration files and databases. In many cases, the encryption can be far-reaching and include all files for normal operation of the company, such as emails and documents. Once the victim has been compromised, they must pay a ransom to the attacker to receive a key to decrypt all the files back to their normal state. Unfortunately, some ransomware attacks may result in permanent damage. Even when trying to restore from backup, the files will continue to be encrypted, rendering normal business continuity methods ineffective.

There is a third form of attack on availability (as well as data integrity) which is worth touching upon. Some hackers are happy to disrupt a website by posting different content onto a page. Sometimes this can be for political reasons, such as the hacker group Anonymous sabotaging websites of specific organisations by replacing normal content with political messages or offensive images. These are designed to embarrass the victim as well as impact site availability.

Defending against DDoS and ransomware attacks is complex. For the former, there are multiple controls that a DevOps engineering team would need to consider for application code and the environments in which they are hosted. These include using good coding principles such as validation, memory management and application code flow, and good operations principles such as load balancing, firewalls and resource management. We will delve into some of these later in Chapter 5.

What about trust?

In the context of the CIA triad, *trust* is an overarching requirement affecting all three elements. Being able to identify resources and allow authorised actions from authenticated users and services is a fundamental component of security. One of the key objectives of a hacker is to obtain the details of a trusted resource, allowing them to jump across security domains to reach their target. Hackers use several methods, including cracking passwords of credentials with elevated privileges and spoofing certificates to create a trusted connection with a target endpoint. We also see trust being exploited in social engineering and phishing attacks which fool users into parting with useful information such as their credentials. Another key objective of the hacker is to cover their tracks when using their own credentials.

They deny their involvement in malicious activities – a concept known as non-repudiation.

Final thoughts on CIA

Attackers may target the confidentiality, integrity and availability triad for financial gain, to disrupt a service or to make political statements. Many attacks exploit weaknesses in more than one area of the CIA triad. For example, if the integrity of a file is not checked, it may be a Trojan that initiates the ransomware attack that affects availability. Likewise, a DDoS attack may be a cover for an attack to exfiltrate information, targeting availability and confidentiality. While focusing on the triad components forms the basis of your defence strategy, it is also important to factor in protecting trust within your organisation's security policy.

Now that you have an understanding of what types of attacks to be aware of, I will explain who your adversaries are.

Adversaries and their weapons

In *The Art of War*, Sun Tzu states that knowing your enemy and knowing yourself are key strategies for victory. Therefore, identifying your organisation's adversaries is key to building a strong defence against attacks targeting it. As we have seen from the CIA triad above, there are three lines of defence to protect the applications and services your DevOps engineers develop and support. There are also many types of adversaries who target your organisation using a host of techniques.

Insider threat

One significant risk to an organisation involves employees deliberately sabotaging applications and systems. There are multiple reasons behind insider threats, such as termination of employment, seemingly unjustified poor performance reviews, and employees turned rogue or working for outsiders (so called 'disgruntled employees'). These types of attacks are difficult to defend against mainly because employees have a familiarity with the organisation's IT systems, including their potential weaknesses and ways to hide their malicious activities until it is too late. It is recommended practice to follow some basic rules if an employee's contract needs to be terminated. These include escorting the individual off the premises, removing access to all systems immediately, removing any hardware provided by the organisation (such as laptops, mobile phones and hardware tokens), and reclaiming and deactivating all building passes.

DevOps promotes the use of telemetry as a means to provide transparency within a system. The data that these metrics produce allow the monitoring of systems which can detect suspicious behaviour indicative of an insider threat. For example, it should be possible to detect if an employee has changed administrative passwords or is attempting to promote code to production outside of a release cycle. Unit test coverage should be monitored because an inside actor may also attempt to remove unit tests covering security features so that vulnerabilities remain undetected when the code is promoted to production during a normal release cycle. After leaving the company, the employee could exploit these weaknesses to cause further damage. Likewise, regular code reviews, which we will discuss in Chapter 4, can detect backdoors into an application or system created by an internal attacker.

Some insider threats are not intentional on the part of employees; they may, for example, fall victim to a social engineering attack to reveal sensitive details such as their employee credentials to an external attacker. Educating staff is an effective control to identify potential social engineering attacks – particularly phishing attacks, where an email recipient is tricked into opening an attached document containing malware or clicking an embedded link to a malicious website.

Another common problem that may be considered an insider threat is 'fat finger syndrome', in which a member of staff accidentally misconfigures part of a system by manually changing a value. This could have unexpected outcomes such as a system failing or accidental exposure of data on publicly accessible data stores. The role of DevOps in mitigating against this type of threat cannot be understated. When following the three ways of DevOps, all changes to production systems are managed exclusively through automation.

The only changes that can be made to a production system are through application or infrastructure code changes that have been peer reviewed and checked into a source code repository by an engineer and that successfully completed automated testing. Unlike manual changes, where mistakes are difficult to find and fix, especially if the change is not documented, automated changes can be easily rolled back if they have a detrimental effect on the live system.

Malicious hacker

A malicious hacker is described in the US Department of Commerce National Institute of Standards and Technology (NIST)'s Special Publication 'An Introduction to Information Security (SP 800-12r1)', authored by Michael Nieles et al, as 'an individual or group who use an understanding of systems, networking, and programming to illegally access systems, cause damage, or steal information'. There are several types of malicious hackers, each with a different motivation and *modus operandi*. At one end of the spectrum are government-backed state actors, who use cybersecurity as weapons against other states, top-ranking individuals or private entities involved in state supply chains, and at the other end are those who hack solely for the thrill of compromising a system. Between these extremities, there are organised criminal or terrorist groups who are motivated by financial gain and disruption to services; industrial spies who seek to gain competitive advantages over rival organisations by stealing intellectual property; and spammers, phishers and botnet operators, which often work collectively to steal identities, plant malicious code or illicit financial gain.

Hackers use various different techniques, often chained together, to meet their objectives. Once they have gained access to your network, they will try to find a route to the target machine. Starting with an easily exploitable vulnerability, they attempt to elevate their privileges to access other machines on the network. The process of attacking one machine to access another is called *pivoting*. As they compromise each machine, attackers install malware, exfiltrate useful files, run scans and look for vulnerabilities to establish a path to their goal. Some organisations may be under attack for several days, weeks or even months before the attacker administers the *coup de grâce*. DevOps practices and security hygiene should identify and stop these types of attacks to build a more secure environment for your customers.

There are many controls that mitigate hacks by these malicious actors. Engineers within a DevOps team need to defend in depth to protect their applications and services. In other words, they need to provide comprehensive protection of their application and infrastructure by implementing plenty of controls. As I will discuss in later chapters, the three layers of DevOps security offer solutions for defending in depth against malicious hackers; ensuring staff are educated engineers following good coding principles, and security vulnerabilities are quickly identified and fixed.

Weapons of a hacker

Hackers are equipped with an arsenal of weapons to attack systems with varying degrees of sophistication. Their tools can be classified under four categories, which form the four basic phases of an attack:

1. Reconnaissance
2. Scanning
3. Exploitation and weaponisation
4. Control and management

The objective of reconnaissance is to learn as much about the target organisation as possible to build a picture of the company's internet profile, its employees, partners and any other information that an attacker may find useful. This activity, known as open source intelligence (OSINT), normally involves using publicly available sources such as social media, search engines, Whois (a record of registered users of internet resources such as domain names and IP addresses), and social engineering to gather information. Since much of this information is gathered from public sources, the bulk of this activity is low-key and carried out without the knowledge of the target. There are various controls that mitigate this activity, the key one being a monitoring and alerting system, which we will discuss in Chapter 6. A robust education strategy will improve the security culture of your organisation by teaching employees to be wary of what they post on social media, how they answer unsolicited calls, and how they spot and deal with potential phishing emails.

The information gleaned from the reconnaissance phase may provide enough information for the attacker to employ more intrusive tools such as network scanning and port scanning utilities. These identify the IP addresses the target uses and any open ports within those IP addresses. Attackers will also scan these IP addresses and ports using vulnerability scanning tools that identify applications and services, their versions and any known vulnerabilities associated with them. Many of these known vulnerabilities are often found in older, unpatched versions of software.

Once the hacker has an understanding of the network and types of vulnerable applications and services hosted on these platforms, the next two phases allow the attacker to exploit the known weaknesses to meet their final objective and cover their tracks. This is where the tools become more sophisticated. The goal in this phase is to deploy weaponised utilities, such as viruses, worms and Trojan horses, on the compromised server. These tools are used for many scenarios including cracking admin passwords, harvesting information, and creating open channels for command and control activities. They may also manipulate log files to obfuscate their activities. Hackers often use open source tools specifically designed for these tasks, many of which are available in a Linux distribution ('distro') called Kali. Hackers using this distro have access to a whole suite of open source tools that their peers have developed specially to exploit gaps in an organisation's defences. Patching and monitoring are effective controls against many of these tools; however, a determined hacker can be creative in finding weaknesses in your applications and services hosted in your environment. Your goal is to stay one step ahead of these attackers to reduce the risk of becoming a victim.

DevOps practices, when applied correctly, are well equipped to defend against common vulnerabilities within your organisation. Correct application entails having the capability to identify and patch software to a more secure version as quickly as possible. Not only should this apply to the software that engineers develop in-house, it should also apply to third-party dependencies used in their own applications and the environment on which these applications are hosted. Using DevOps, patched versions of software and infrastructure can be deployed with minimal disruption to the value stream as an important line of defence. However, because the number of vulnerabilities grows daily, this should be embedded into the workflow of each value stream.

Unlike known vulnerabilities, a zero-day vulnerability is one that is unknown to both consumers and providers of the affected software until it is actively exploited. For these vulnerabilities, a patch may not be available for several days or weeks, after the vulnerability is made public. DevOps teams may need to introduce other controls to mitigate against zero-day vulnerabilities such as replacing an affected library with a more secure one from another third party or removing an affected library's functionality until a patch is made available. The ability of DevOps teams to act quickly to address a zero-day vulnerability while minimising the effect on customers is important. Through DevOps, engineers are enabled to develop, test and deploy a secure version of the application or service in a short time frame, reducing the likelihood that the vulnerability will be exploited. Later in this book, I will examine in more detail how to identify and fix common vulnerabilities and zero-day vulnerabilities as part of the DevOps way of working.

Although hackers have a number of weapons at their disposal, there are various layers of controls to defend against most of them. In the next chapter, I will describe how to explicitly embed security practices into the DevOps way of working to help an organisation to protect their valuable digital assets. As has been demonstrated, DevOps intrinsically provides some natural defences against common exploits, but these don't have the depth required to provide a more robust fortress. There must be a collective and conscious effort to embed security practices into the DevOps way of working.

Conclusion

By now you should have the basic understanding of what security means to your organisation. There are many individuals, groups or state actors determined to exploit your organisation for personal, financial or political gain. They look to expose confidential information, diminish the integrity of an organisation or affect the availability of the services needed to run an organisation smoothly. Whatever their motives, they are often persistent; the greater the reward, the greater the effort to find a route to the jewels of an organisation. Even if your organisation is not the ultimate target of hackers, they may exploit weaknesses in your applications and services, leaving a trail of collateral and reputational damage in their wake. The hacker community is well organised and has developed a large number of effective tools over many years to simplify their tasks. Your threats are not just external but internal too. It is the collective responsibility of everyone within the DevOps organisation to protect your assets and those of your customers.

THREE

DevSecOps

'DevSecOps', a combination of the first letters of 'development', 'security' and 'operations', implies the merging of developer processes and operations processes (DevOps) with security processes. In reality, DevSecOps is an extension of the cultural shift towards DevOps incorporating everyone within the organisation, including business stakeholders, security, architects, product owners, and development and operations engineers. DevSecOps has become the more established term used within the industry, although you may come across references to SecDevOps or DevOpsSec or even just SecOps. For the rest of this book, I will use the term DevSecOps.

From a technical perspective, DevOps is the collaboration of personnel and processes from development and operations to form a single Agile delivery team consisting of stakeholders, customers, engineers and testers. This team works together to deliver software and infrastructure at pace via an automated CI/CD pipeline. It is natural to assume that DevSecOps is an extension of this and merges *security engineers* with DevOps into a single team. Unfortunately, this model is far too simple to describe the role of security within DevOps. Security is concocted of a multi-disciplined group of individuals, each with their own specific role.

How do you determine which security discipline is required in DevSecOps? Even when you have established which type of security role is needed in DevOps, there are rarely enough security engineers in an organisation to assign to each DevOps team. How is it possible to share so few resources across the many small, cross-functional teams working on the various value streams? According to Gene Kim et al, security staff are outnumbered by developers at a ratio of 100 to 1. Even if you are fortunate enough to have a quota of security engineers at your disposal within your organisation, they are each likely to have a specific security function which is unlikely to cover all the security requirements of the DevOps team. You would want the security engineer to engage with the cross-functional DevOps team using many different skills. Are they more effective doing code reviews? Fixing security defects? Writing security tests? Performing security risk assessments? Configuring networks and firewall rules? Implementing access controls? The chances are you are asking too much of your security engineer: not only do you need a multi-skilled individual, you are also asking them to spread themselves too thin across multiple teams, diminishing their effectiveness considerably.

On the other hand, if security remains outside the DevOps function, they will continue to perform the gateway role and sign off deliverables once the DevOps teams have completed their work. This practice is counterintuitive to the DevOps principles outlined in Chapter 1; as a result, the DevOps teams may look to bypass established security controls within the delivery pipeline to maintain the flow of the value streams.

This analysis leads me to conclude that DevSecOps should not be defined as a team consisting of development, operations and security roles. In fact, I believe that DevSecOps is a term that describes a cross-functional DevOps team that integrates security practices within their own processes to deliver *secure* software and infrastructure. To put it another way, DevSecOps is DevOps done securely. In the words of Eliza-May Austin, 'DevSecOps teams simply don't exist.'

Security implied in DevOps

Security is often described as an implicit feature of DevOps. This is based on the notion that DevOps is secure because DevOps practices, such as code reviews, automated testing, automated deployments, 'blue-green' or 'a/b' testing, and small iterative changes to production code are security controls in their own right. The argument goes that if DevOps teams follow

DevOps processes then they can be confident that the quality and security of the end product is better than if they did not follow DevOps processes. However, I believe leaders working with delivery teams adopting DevOps practices need to challenge this point of view. Security is not the product of DevOps practices, although they help. In reality, security needs to be addressed explicitly by DevOps teams to ensure that the applications and services they produce are not at risk of being compromised by a malicious actor. This means that DevOps teams need to adopt a security mindset and apply security principles to their day-to-day activities. As mentioned above, it is not as simple as including an infosec engineer in a DevOps team.

It is dangerous to assume that following DevOps practices leads to more secure software. If your organisation has yet to adopt DevOps or has just adopted DevOps, the temptation is to state your case with security professionals that there is a lower risk of security incidents in DevOps. You may stretch this argument further to state that automated deployments and continuous delivery in DevOps allows for a faster recovery from security incidents than traditional methodologies. In reality, security teams are more likely to take a closer look at the way your team is working and enforce practices that are counter-productive, creating bottlenecks that reduce your teams' ability to deliver new features and services to their customers as quickly as your business stakeholders would like. Worse still, the security departments may block releases if they believe that there is an increased risk of cyber-attacks to the company. Ultimately, this may create the need to bypass security assessments completely, resulting in vulnerable software being delivered to your customers. Adherence to a robust security policy is as important to DevOps as it is to any other delivery methodology.

Points of contention between DevOps and security teams

The role of security teams in software development is an important one. Traditionally, they have played a pivotal role in making sure architects, developers, network engineers and project managers factor in security when designing, developing and deploying application software. The typical methodology involves a security architect reviewing the security of the proposed solution. Then, when development is completed, security engineers perform code reviews and security testers run penetration tests. During the software development lifecycle, network security engineers will be called upon to make firewall configuration changes to support development, testing and ultimately the running of the application in production. In addition, other security functions set up access policies and assign users to roles or provide encryption keys or certificates. Each of these tasks requires a lead time and detailed documentation to be carried out successfully. This way of engaging with security, although not ideal, was accepted by most companies as the right way; it complements the waterfall methodology in which applications go through a linear process of design, development, testing, configuration and deployment.

But the role of security does not stop there. When the application is running in production, the security operations centre (SOC) is responsible for monitoring live systems and ensuring operations engineers fix any security issues that occur using the documentation provided by the project delivery team for reference. This documentation would include architectural diagrams, application interfaces, network topology and database schemas. Often, by this point the project delivery team has been disbanded, so the documentation is all they have to work with. If they are lucky, some developers may be available to help decipher their own code to understand what is causing the problem, but in many cases the code is so old that programmers have forgotten how the code works or what an algorithm is meant to do.

As discussed earlier, DevOps pulls together development and operations teams to apply Agile methodologies to design, develop and deploy applications and infrastructure using automated processes, continuous integration and short feedback loops. The DevOps team is self-sufficient: developers and operations engineers can make small incremental changes in application or infrastructure code to enhance a product's existing features or introduce new ones. The DevOps process allows for multiple releases in a short space of time, and a mature DevOps function can deploy to a production environment multiple times per day. Even less mature teams may deploy as frequently as once a week or fortnightly. Unfortunately, this cadence does not work well with the traditional security processes described above. For example, DevOps builds on Agile, which downplays the need for fully documented systems required by security teams; it values working software over comprehensive documentation. Therefore, security architects are unable to review a complete architecture before any programming starts, network engineers are unable to see a topology of the network until later on in the process and the SOC team has nothing to work with other than 'working software'. Code reviews are normally carried out when development is complete, but in DevOps, which is based on product value streams, development is a continuous process that never ends. This means that security engineers are rarely asked to review source code, or they are asked to review source code at an unsustainably high frequency. As developers and operations engineers work on their code, there is often an instant requirement for a firewall change or a new user role or access policy, yet security teams require complex forms to be completed and approvals from senior management before changes can be made.

Furthermore, security teams often need to support many different functions within an organisation, not just software delivery. In my experience working in large fintech organisations, their security team methodologies tend to support the wider ecosystem far removed from DevOps practices.

This battle between traditional security practices and DevOps can lead to distrust. DevOps engineers are reluctant to engage with security teams for fear of slowing the progress of their work, and security teams are reluctant to allow deployments if they cannot verify their security.

A layered approach to effective DevSecOps

Although cross-functional DevOps teams should bring members of the infosec department on board, in reality this is not feasible. It simply doesn't scale. We need another approach to avoid an escalation in the tension between security functions and DevOps teams. In this book I will show you a strategy that gives regulators, stakeholders and your customers confidence that you are delivering new live and experimental features at the pace necessary to remain competitive without compromising security. My solution is a layered approach incorporating the main security ingredients to form a solid foundation for adopting DevOps within your organisation. This approach enables DevOps teams to incorporate security into their daily activities while giving the under-resourced security departments time to focus on a strategy that supports DevOps. These layers can be adopted individually, but the real benefits come from putting all the layers in place: the whole being more valuable than the sum of its parts. In this section, I introduce the layers and provide a brief overview of each.

Three layers overview

Security is a broad topic. There are many fields covering security, such as data security, cryptography, network security and physical security. Within these separate fields, there are different subjects, each with a plethora of information; therefore, defining security within a particular way of working can be daunting for the teams working on delivering secure products to their customers. My *three layers* help DevOps teams to incorporate security while remaining focused on the functional requirements of the product they are developing and delivering.

These layers are the foundations on which DevSecOps should be built. Security should never be an afterthought; it must be integrated into every aspect of product delivery. However, the challenge to product owners is to deliver new features and products to market as quickly as possible in order to remain competitive. As we saw in Chapter 1, DevOps evolved from the Agile delivery movement, which offers a greater cadence to delivery than traditional waterfall methodologies. Teams of software engineers, network engineers, testers, architects and product owners work in short cycles to deploy incremental changes to a product or service while minimising or negating the impact on the customer. In this model, security becomes an afterthought – a penetration test may be scheduled, or a few automated security tools are hastily integrated into the CI/CD pipeline. DevOps teams are often reliant on separate security teams that are not part of the delivery process and, therefore, become bottlenecks affecting the DevOps flow. As a result, the role of security is to monitor for and react to security incidents after a product has gone live, rather than helping teams develop secure software. In many cases, there are no short feedback loops to address problems as they arise, and this often leads to prolonged outages or major disruption to customers.

The solution is not to try to integrate out-of-date security practices but to rebuild security practices from the ground up to support DevOps teams. Without the *three layers* of DevSecOps, you only have a DevOps team plus an ineffective security 'bolt-on'.

The first layer is *education* – or, more specifically, security education. Everyone involved in DevOps should know how security affects what they do, including how it is involved in creating new features and products, designing applications, writing code, testing features, deploying code, building the infrastructure, identifying and responding to incidents, or selling to customers. Security must play a fundamental role in what we do. Your cross-functional teams should be able to operate every day with the curiosity needed to challenge their peers about insecure practices and show them the right way to bring security into their work. Likewise, everyone should have the resources and motivation to put security at the heart of everything they do.

The second layer builds on the first by ensuring that good design principles are followed at all times. *Secure by design* means putting security at the forefront of what you do, not only in the way you design and code your products but also in how you design the assets you need to deliver your products and how you plan the processes and interactions between individuals, teams and customers. Good design principles are not necessarily explicitly based on security; they're based on writing clean code to reduce the number of defects produced and designing clean architecture to create robust products and processes that support those products. 'Secure by design' is not exclusively focused on technical design – it also incorporates the controls you put in place through the day-to-day activities of your teams.

Finally, you need to demonstrate that your teams are applying their security knowledge and implementing good-quality design principles through the processes of *automation*. This forms the third layer of our foundation for implementing robust security measures within your DevOps ecosystem. Automation is at the core of DevOps, and automated security testing needs to be embedded within this core. It is essential, however, that automated testing is built upon the first two layers.

Without the solid foundations of education and strong design principles, you will not be able to reap the benefits of a strong and effective testing strategy.

The three layers of DevSecOps

In the next three chapters, I describe how to build each of the three layers to empower DevOps teams to construct secure applications using processes that are equally secure. Together, these three layers describe DevSecOps – not as a team, but as a culture. It is important to not skip the first two chapters and race to the automated testing chapter. Like any reliable structure, each layer needs to have time to settle. Equally, though, it's not necessary to plan and build the whole of each layer before moving on to the next. In fact, the principles of Agile should apply to building these layers, meaning you should adopt an incremental approach to build vertical slices that allow your teams to continuously improve how individuals and teams are educated, apply their skills and assess progress.

Conclusion

In this chapter, I introduced the concept of DevSecOps, the need for it to be explicitly defined and explained, and the points of contention that exist between traditional cybersecurity ways of working and DevOps practices.

This book proposes a layered approach to solving this problem, enabling DevOps teams to deliver applications and services securely, while allowing security departments to focus on strategy to support the business needs for faster digital delivery to their demanding customers.

I have introduced the three layers that sit below DevOps to form the foundation of DevSecOps. Each layer builds on the previous one to create strata of security education, secure by design and security test automation. In the next chapter, I explain the details of layer 1.

FOUR

Layer 1: Security Education

In this chapter, I develop the first of the three layers by explaining why security education is so important to DevOps before describing the various ways in which education can be delivered within your organisation to promote a strong understanding of security. I also explain how you should identify those with a deeper interest in security within the engineering teams so they can play an important role in expanding the security skills of DevOps teams. Finally, I highlight the importance of maintaining an elevated level of security knowledge and skills within your organisation to continue offering high-quality products to your customers.

Importance of security education

As the saying goes: knowledge is power. Understanding how an attacker may exploit weaknesses in your application or infrastructure enables engineers to think about more secure ways of writing code. Although adopting good design principles is a partial solution to the problem of vulnerable code, engineers are often ignorant of why writing code in a particular way is inadvisable. While they may know that using immutable objects is an acceptable design practice, they may not understand the importance of following this practice from a security perspective. Likewise, if inconsistent validation rules are written within applications, they become too difficult to manage, and the developer may choose a path of least resistance and produce less effective validation.

If DevOps engineers know what to do, but not why or even how to do it, then they are less likely to implement a particular logic or algorithm, even though it is the most secure method. As such, it is important the developers learn the basics of writing secure code and understand the implications of writing insecure code. Likewise, operations engineers must be proficient in maintaining the secure environments in which the developers work, whether this is the CI/CD pipeline or the infrastructure on which the software is deployed. SANS has long maintained that one of the primary causes of computer security vulnerability is 'assigning untrained people to maintain security and providing neither the training nor the time to make it possible to learn and do the job'.

Education is important for all roles that involve a level of technical ability. Managers expect their engineers to have a deep understanding of the latest (and greatest) tools available to them, such as modern frameworks, newest development languages and the multitude of cloud solutions. Managers are often happy to factor in new skills as part of a team's target, especially if they give their products a competitive edge or a more cost-efficient (ie faster) way of adding new features. Some courses, such as security training, are often regarded as a necessary evil: something we all have to do to satisfy the requirements of regulators or compliance. In our ever-changing world, where new technologies are released frequently and new attack vectors expose more sophisticated vulnerabilities of these technologies, continuous learning is essential for the ongoing commercial success of an organisation. Without a strategy for learning, organisations risk being left behind their competitors – or, worse still, risk being attacked by cyber criminals.

In the UK, and in many other countries, the education system is prescriptive, repetitive and one way. In the classrooms, a number of students listen to a teacher who imparts knowledge in accordance with the requirements of a curriculum. This format normally continues into the workplace, where external or internal instructors teach a number of employees a new skill. Often, this is automated via a series of videos and online tutorials, followed by a test that demonstrates the student's understanding of the topics being taught. The problem with this process is that not all students gain knowledge that is useful beyond the scope of a test, an exam or even the course they have just attended.

In order for education to be effective, your employees must have a wide range of options.

Ray Coulstock has been providing technical training all around the world for over two decades. He notes that there are many cultural factors involved in how people respond to education. In some countries, training tends to be more prescriptive, giving students step-by-step instructions to develop their understanding of a concept; in other countries, students are more proactive and can grow an idea from the seeds planted by the trainer. Coulstock also points out that in addition to cultural differences, individuals respond differently based on their own personal trends. Some people respond to audio stimulation, some respond to visual stimulation and some respond to written stimulation; a few respond best to all three. Ray's experience has also taught him that training needs to be part of a wider training plan tailored to the individual. He adds, 'It's no use sending someone on a course if they are not going to use their new skill in their day job from day one.' The plan should be based around the specific requirement to meet a business need. For example, if the engineers are developing a new security feature requiring multi-factor authentication, they will need to learn what that entails to ensure they build a secure solution. The engineer should be given the opportunity to learn the skills required and then apply those skills to the products being worked on. In the meantime, the engineer should be given the opportunity to work with their peers, which allows them to pass on the skill.

You should encourage team members within the organisation to identify the learning format that suits them best as well as the topics that will improve their ability to deliver value to your customers. As a leader, it is your responsibility to give the staff the funds to follow the learning platform that benefits them the most. This will benefit your customers through well-designed, reliable and secure products and services that they will enjoy using.

It's important to note that timing is critical since much of what is learned can be forgotten quickly. Therefore, it is important that training coincides with a technical or business requirement; for example, when you introduce a new security tool into the DevOps process or when there is a drive to reduce the number of incidents relating to a specific vulnerability that engineers need to learn to identify and fix.

Agile and DevOps promote the practices of continuous integration, continuous delivery and continuous deployment. Added to this is continuous education. With each release, there is an opportunity to learn about customers' engagement with the products and services your organisation develops. Furthermore, there is opportunity to educate everyone involved in the value stream on security-related functional and non-functional requirements. In this chapter, I describe the education methodologies used to increase security knowledge within the organisation. But first, I would like to highlight a group of people who will play a pivotal role within DevSecOps.

Security champions

Everyone responds differently to education. Some learn just enough to do their job well, others learn more than is needed, and a few will struggle to grasp certain concepts required to complete their tasks effectively. This is a normal distribution within an organisation. Occasionally, there are individuals who learn a new discipline that sparks a passion within them. In my experience, it is important to feed this enthusiasm by giving the individual the time and resources to grow. With the industry struggling to find enough quality security resources to support the increasing scale of software engineering, finding and helping members of your organisation who are prepared to pivot their career towards security is crucial. As they increase their security knowledge and skills, in addition to their existing skills, they will become key players in helping DevOps teams improve the security of the products they create. We call these people *security champions*. They play an important role in the continuing education of other team members by supporting peer reviews, engaging in threat modelling exercises and helping peers analyse outputs from security scanning tools. The key advantage security champions have over a central cybersecurity function is having specialist skills (such as software development, architecture and configuration management) in addition to their growing security knowledge.

As you introduce different education tools to improve security within the delivery lifecycle, it is important to identify the security champions early to continue investing in them. Their passion for security will improve the overall security of the products they are helping to deliver. You should allow them to work as cross-cutting members of the organisation, helping multiple delivery teams to improve the security of the products they are working on. There is a temptation to place them within the teams that are working on the most valuable products in terms of revenue, but you should avoid siloing your best individuals in this way. By giving security champions a wider remit, they can potentially nurture aspiring security champions across the organisation's product portfolio by becoming security evangelists.

Security champions that evolve from being software or infrastructure engineers also play a crucial role in building security into the development and automation process. They become prime candidates for writing re-usable security components that other developers can consume when writing software. They are also well-placed to write security tests that can be integrated into the CI/CD pipeline or configure the tools used for testing the security of your products.

Security champions are not security experts, so you should not expect them to replace the entire function of a security team. However, their role can provide a conduit between your organisation's central cybersecurity specialists and the DevOps engineering specialists to improve the relationship between the two groups, providing better outcomes for your customers.

On the few occasions I have come across security engineers who have pivoted the other way, moving from a central security role to a development role, they have played a similar role to the security champion. Their main strengths lay in the wealth of security experience they bring to a DevOps team and their potential to motivate your next security champions.

Through education, you will discover the engineers inspired to further develop their own security knowledge and skills. As these individuals grow in maturity and confidence, they too can educate and inspire DevOps engineers to become better at doing security and doing it well.

Gamified learning

Education should be fun. The more candidates enjoy the learning process, the more they are likely to remain engaged with the subject, which, in turn, increases their potential to absorb more knowledge. Gamification is the act of turning something that is not a game into a game that involves scoring points or competing against others; however, gamification in the context of learning is more complex. The aim is to increase enjoyment, engagement and loyalty in performing a set of tasks that ultimately enhance the learner's capabilities and knowledge. There are many vendors who offer gamification services; their games all assume a basic understanding of coding practices and may offer modules specific to the language that is familiar to the engineers. These include application development languages such as Java and .NET, as well as infrastructure coding languages such as Terraform and Ansible.

Usually, players are presented with a detailed description of a particular type of vulnerability (or common weakness) and an integrated development environment (IDE) emulator containing code with an example of the vulnerability. These tools teach players how to directly exploit weaknesses deliberately written into the code. Hacking code in a safe environment is a fun way for engineers to learn how poorly written code can have severe consequences. To score points in the game, the engineer has to fix the vulnerability. There are normally added incentives for the participants such as a certification programme or tournaments. Certification programmes provide engineers with badges of honour, which enhance their profile among their peers or improve their self-worth. Tournaments add intense competition between individuals and teams vying to prove their statuses within their organisation as leaders in security. With each successive tournament, they need to defend their positions, incentivising them to learn more about writing secure code. Matias Madou, co-founder and CTO of Secure Code Warrior, offers a cautionary note on running leagues and tournaments: 'Don't focus exclusively on the leaders. You need to focus on the lower ranked individuals in order to keep them engaged. You don't want to lose them.'

It is not essential to use third-party gamification

tools. It is possible to develop internally hosted games by collating and presenting real-time data to the engineers. The goal is not necessarily education; rather, it encourages safer and more secure products and services. However, education plays a part in understanding how to tackle certain types of vulnerabilities. For example, a dashboard could indicate the number of vulnerabilities identified and fixed by each engineer over a specific period of time. Although this particular scenario requires a reasonably high level of maturity within the team – not only to identify and fix issues but also in having the metrics to present to them in a dashboard – gamification in real time can be adopted in small increments. The first step could be to collect information about the adoption of security testing tools, pitting teams against each other to incorporate security testing into their working practices. Over time, data relating to the number of vulnerabilities these tools identify can be measured before recording the number and type that are fixed and by which engineers. This form of gamification creates competition between teams and individuals to reduce the number of vulnerabilities being promoted to a live environment, which makes the products safer for their customers.

Gamification, if done correctly, can have high engagement among engineers. Debasis Das, of one of the world's largest banking organisations, has implemented a security gamification product across the company's digital organisation in India. He identifies a number of key factors that led to the widespread use of gamification among his DevOps teams. This high uptake has significantly reduced the number of vulnerabilities introduced into the software solutions that the DevOps teams produce. Running regular tournaments has been highly successful, but Debasis cautions that this must be done correctly to avoid lethargy among engineers. A successful tournament must provide incentives beyond bragging rights; therefore, he suggests that teams and individuals play for prizes that are worth playing for. If the prizes are of value to the engineers, they are more likely to want to be involved. He also advises that the organisation's leadership team must fully support tournaments to avoid conflicts of priorities between delivery and engineering. Delivery leads must be open to allowing development to stop for at least half a day for the tournament to go ahead. Debasis says of the senior leadership team: 'They should agree on a time and block out the calendars of the engineers to participate in the tournament.' Leadership teams must also support the ongoing education of the engineers by allowing a percentage of the engineering workload to be dedicated to gamification (as well as other forms of training and learning) – not just in security, but in all areas that improve the quality of the end products.

Self-paced gamification versus tournaments

Gamification *players* are normally able to learn at their own pace, but they must be given tools that pique their interest. When browsing the top games in the various application stores (such as Apple's App Store and Google's Play Store), there is a plethora of games that have addictive qualities. Successful games are easy to use, have multiple levels which become progressively more difficult to complete, feature rewards (sometimes known as treasure chests) and offer opportunities to share success with peers. Learning security through gamification provides a similar experience, motivating players to navigate through a series of progressively more complex vulnerabilities to identify, exploit and fix. After reaching certain milestones, the tools may reward the user with a badge or certificate which can be shared on social media. They often come with built-in league tables that allow individuals to compare their progress. This can act as an incentive for engineers, although you must avoid apathy among those who do not fare so well.

Despite its advantages, this self-paced approach is not suitable for everyone. Some engineers may be restricted by time constraints or other factors that slow or halt their progress. In this case, it is worth running a security coding tournament to provide fun organised training. Tournaments can be run between multiple teams or individuals, with prizes awarded to the top performers. Although a tournament may involve a large proportion of the engineering workforce over an extended period of time, the rewards are evident in the improved code quality and the potential reduction in vulnerabilities being discovered in a production environment. It is well known that the cost of fixing bugs in production is more expensive than fixing them earlier in the lifecycle, and an educated and enthused engineer is more likely to identify and fix issues during development rather than after the application has been deployed to a live environment.

One of the potential pitfalls of gamification is that, in order to gain a competitive edge, players may choose to excel in something they already know rather than challenge themselves on learning new material. Encourage your engineers to push themselves to learn new skills and try new modules within the gamification platform. The number of modules completed and their comparative difficulty rating could be incorporated into dashboards and used as a measure of success for individuals and teams.

Instructor-led training

There are advocates who argue that traditional teacher-led training in a workplace should be avoided. However, I believe that businesses can benefit from teacher-led security training applied in the right way. The main advantage is that the course content can be tailored to the needs of the business: the content can focus on certain types of security weaknesses that have been identified during vulnerability assessments, or the course can be contextualised with examples from the organisation's source code repository. In schools, the teachers follow a curriculum which presents course material to the students in a predetermined order, irrespective of the students' motivation to learn. In business, you should avoid the temptation to follow a pre-set syllabus if the content is not aligned to your organisation's requirements, especially if the course is fixed by the trainer without any input from the candidates. Remember that you are not dealing with schoolteachers but with professional instructors who provide a service to you; you can design the syllabus that meets your organisation's requirements.

With any instructor-led training, it is imperative you ask the supplier to sign a non-disclosure agreement and limit access to the resources required to perform the training. If an access key is required to enter the facilities, ensure that access is time limited and is only for the training room. If the trainer needs to access digital resources on the company network, implement policies to restrict access to the absolute minimum to the relevant information. The instructor may wish to plug devices, such as USB sticks, removable media devices, or laptops and tablets, into your network, per-haps to share documents with the course attendees. To protect your company from malware or viruses being accidentally (or even deliberately) installed by the instructor, you must follow your organisation's security policies in relation to BYODs (bring your own devices).

The human element of instructor-led training is an important factor, as it provides an opportunity for stu-dents to ask specific questions to drill into a topic or to clarify a point. In business, an effective instructor may determine from the students' feedback that pivoting the content during the course will produce the best outcome for the students and the organisation. Students and instructors can develop a strong rapport, which can create an ideal learning environment, resulting in your DevOps engineers gaining a solid understanding of security in the context of what they are doing via a two-way conversation with the instructor.

Cost is often a prohibitive factor for instructor-led training. However, online training can mitigate this as well as broadening the target audience and reducing travel costs for the trainer and the attendees. Remote video conferencing tools such as Zoom, Microsoft Teams and Google Hangouts make this approach easier.

Instructor-led classroom training

Security training sessions that take place in a classroom normally target a small audience of individuals and usually take place over two or three days. The main advantage is the intimacy and trust that the instructor can build with the students. Although any trainer will allow each attendee to give a short introduction within the formal bounds of a lesson, an effective trainer will meet the students informally and start to build a relationship before the class starts. Even more introverted students can be included in the discussions that play out in the classroom.

Relationships are often built during short coffee breaks or during lunch. With this in mind, engineers attending security training classes should refrain from going back to work during breaks, to answer calls, respond to emails or catch up on the latest news with their peers.

One of the benefits of classroom training is the opportunity to have the instructor join the DevOps teams during their daily activities to see first-hand the issues that the training course needs to address. Once in the classroom, the instructor can work directly with the application software and infrastructure code to identify and address the most common types of vulnerabilities identified within your team's source code. In this case, you would normally ask the trainer to sign a non-disclosure agreement, and you should ask the instructor to leave all training materials with the organisation.

Classroom training benefits from taking place in a closed room that provides a safe environment for confidential discussions. The openness and frankness this environment fosters allow the instructor to focus on the real issues faced by DevOps engineers. The attendees learn how to identify and fix live issues in their products and may learn to identify the most common weaknesses in their products. The ability for the tutor to pivot the course content based on the real demands of the DevOps teams produces better outcomes than less interactive courses.

Instructor-led online training

Instructor-led online training offers similar benefits to classroom-led training in as much that it can be tailored to the specific requirements of the organisation and give attendees an opportunity to interact with the trainer. The main benefit of this particular format is the ability to reach a wider audience in a more cost-effective way. Using online conferencing tools, the trainer and students can share a live video feed so that everyone can see all the other attendees. Indeed, I would suggest that you encourage everyone to turn on their cameras to create a more inclusive environment. Each attendee and the trainer can also present their screens, providing another level of interaction. However, the larger audience and the limitations of using video conferencing technologies generate less intimacy than classroom training. The effect of this is a hierarchical structure within the virtual classroom in which the more confident students are more vocal, while others may hide from view and interact occasionally or not at all. There are some effective tools that the trainer can use to give attendees an opportunity to ask questions. These tools allow individuals to type their questions so that they can be shared with others, voted on and answered by the trainer, or indeed by other more knowledgeable students.

Online training tends to be shorter than classroom training – covering a few hours per day at most. This means that the curriculum needs to be more formal with fewer opportunities to adjust the content as the course progresses. You should agree the content of the course with the trainer and the team before it starts. One way to do this is to ask the DevOps team to work with security to identify which types of vulnerabilities occur more frequently within the organisation; you can discuss the output of this exercise with the trainer and agree on the course content accordingly.

The main disadvantage of the online instructor-led course is the lack of opportunities for the trainer to build relationships with students during breaks: there are no informal chats by the water cooler or over lunch. Therefore, there will be missed opportunities to gain a better understanding of the issues that the course is trying to address.

Online instructor-led training is a popular forum for vendors who provide training on their products such as application security testing tools that your operations engineers need to integrate into the pipeline, or specific development technologies and frameworks that software engineers use. These courses tend to be well-rehearsed by the vendor trainers with little flexibility. However, your DevOps engineers will benefit from seeing experienced subject matter experts using the tools you plan to integrate into the value stream more effectively.

It is worth ensuring that online sessions are recorded so that they can be played back to anyone who wants to learn the relevant topic at a time more suitable to them. Obviously, these recordings will not provide opportunities to ask questions, but they do offer attendees and others the ability to watch them at their own convenience. Fostering a culture of psychological safety must be encouraged within your DevOps teams but, unfortunately, attendees are less likely to air their dirty laundry in recorded sessions, which means that controversial topics are avoided and potential weaknesses in the teams' skills are not addressed.

Instructor-led conference training

Many security and DevOps conferences offer a training stream in addition to the normal conference sessions. These courses last one or two days before or after the conference and are normally led by renowned subject matter experts, authors or leaders in technology. In many cases, these experts actively participate in the conference sessions by giving technical talks, facilitating group sessions or sitting on open discussion panels. Sometimes the training material used during these conference courses covers topics that are at the leading edge of technological advances. This is an important point in the rapidly evolving world of technology – particularly DevSecOps, where the landscape seems to have been in constant flux since the term was first used.

The appeal for most conference course attendees is receiving training from leading industry experts. Many are celebrities in their fields who have published popular books or articles, hosted podcasts or are active on social media. Conference training is an affordable option where bringing the expert charging a high fee to your workplace to train a select few can be prohibitively expensive.

If members of your DevOps teams are attending

a conference that offers a training package, you may want to let highly motivated colleagues join the training sessions for extra in-depth coverage on a topic that interests them. Ultimately, this can benefit your organisation as motivated staff return from their experience to experiment with their new skills, improving processes and flows, leading to better outcomes for your customers.

Conferences are valuable forums for learning in their own right. Not only do attendees benefit from the many talks that take place, they can also learn from the many discussions that take place between sessions when delegates share their own experiences and offer advice to others. It is common for employees who attend these sessions to return to their place of work in a highly motivated state. The key is to correctly harness this increase in enthusiasm. Team members returning from a conference can apply their new security skills to improve the value streams on which they are working, but delegates should also be encouraged to share their new knowledge with their peers within the organisation; although you must avoid returning attendees becoming bottlenecks for the rest of the team. Often, the conference presenters make their material publicly available following each session; this material can be shared with those who did not attend, particularly if the session covers a topic which is relevant to products or technologies being used in the organisation.

Conferences are ideal forums for teaching your employees new skills or new technologies. They are motivational as well as educational. Although it is unlikely your budget extends to everyone who wants to attend, offering it as a reward for those who have demonstrated an aptitude to learning and sharing their knowledge can produce high returns on the investment.

Self-paced learning

Many individuals are self-motivated when it comes to learning something new. They are happy to spend time (often their own time) reading about and playing with new technologies to understand how they may improve the products and services they are developing. There are plenty of resources available to people who like to take this self-paced approach to education. These include the more traditional paper-based resources such as books and magazines (although many are available as e-books or e-magazines); online resources such as blogs, webinars and tutorials; and myriad short video clips available through social media outlets such as YouTube. From an organisational perspective, team members should be learning about topics relevant to the goals of the organisation. This is important in the context of cybersecurity, because individuals may focus on subjects that benefit themselves personally, which may or may not align with the requirements of the organisation. Providing an internal training repository of resources that focus on security can counter this problem. Your organisation's security team can play a key role in identifying the relevant security resources that will benefit the delivery teams. You can build an internal library of books and magazines that cover security as well as the technologies used within the organisation. You can also create an online catalogue of relevant training resources such as links to blogs, videos and articles.

When security is not the core subject of the learning material, the self-paced training resources have a tendency to focus on the technical problems they are trying to solve and do not necessarily emphasise the security elements of the solution. Indeed, it is common for these resources to completely ignore the security aspects of the solution or, at best, provide a disclaimer that the reader should add a security control, but the lack of space or time (or perhaps knowledge) prevents the author from including the control in the text. As an added exercise, your DevOps teams should be encouraged to identify these gaps and document more secure solutions. Your library needs to contain material that specifically covers security and it is equally important to encourage team members to include these resources in their self-paced curriculums.

By providing access, resources and time for your DevOps engineers to educate themselves on security, they are more likely to engage in a self-paced approach to learning. Unlike instructor-led courses, this methodology does not suffer from timing issues. As long as relevant resources are accessible, engineers will use the material to help them overcome problems that are affecting them when they need it. In a DevOps environment, the provision of just-in-time resources to solve problems helps remove blockers and improve the flow of the value stream.

Self-paced learning is also cost effective. Some resources you will have to purchase, such as books and tutorials (although these are cheap compared to instructor-led courses and conference tickets); however, there is also a huge amount of free security education material available online, particularly resources available through OWASP, SANS, MITRE and local government agencies. The following subsections provide a brief overview of the various types of self-paced learning resources and their relevance to security within DevOps.

Tutorials

A popular method of self-paced learning is referencing online tutorials. Tutorials are video clips that provide a step-by-step guide which DevOps engineers can follow in their own test lab. Generally, they are effective at helping engineers remove an obstacle within their workflow; however, security is often an afterthought or is completely missing unless the main topic of the tutorial is security. Your engineers must be made aware of this limitation to avoid the risk of them introducing security vulnerabilities into the application or infrastructure code. On the other hand, many tutorials can be effective in teaching engineers how to code securely, perhaps by showing them how to change insecure default configurations, manage secrets safely or create secure channels between components. Some tutorials are vendor specific, such as cloud providers and security tools. They may offer subscriptions to dedicated video channels containing a multitude of educational material covering a huge number of topics, and your vendors may offer packages in which your engineers subscribe to specific topics within these channels. This prevents engineers from being overwhelmed with content that is not relevant to their needs. Technology vendors may offer video learning courses covering specific topics as part of their own online learning programmes. In some cases, they may offer industry-recognised certification that allows students to earn rewards for successfully completing an online learning programme.

Many tutorials are presented by well-known experts in the field of security and DevOps whose celebrity status make them appealing to engineers. Some experts may be from overseas, so this is the only opportunity for students to learn from their heroes. In DevOps and security, these experts have huge followings and their online content is highly regarded. Working with your teams to make this content available to all team members, particularly subscription-based content, encourages learning from the best.

Books, magazines and whitepapers

Accessibility to books, magazines and whitepapers covering every topic imaginable is embedded in our lifestyle, from online booksellers offering printed and digital books to communities offering printed or online magazines, whitepapers and blogs. Sometimes, the cheapest and easiest way to learn a subject is to read about it and then practise what you have learned.

Over the years, many published books have become 'must-read' additions to people's libraries. These books tend to focus on key principles often based on research carried out over a number of years, or on new ways of working that have become mainstream. During the early Agile movement, Martin Fowler and Kent Beck authored several books that are as relevant today as they were when first published. Robert C Martin published a number of books on designing and developing applications in which many of the principles described nearly two decades ago are cited by developers today as core requirements. Today, much of the architecture we use in modern applications has its roots in the 'Gang of Four' (Gamma, Helm, Johnson and Vlissides) study on *Design Patterns* as well as Eric Evans' *Domain Driven Design*. More recently, Jez Humble, Gene Kim, John Willis and Patrick Debois have documented best practices in adopting DevOps. Organisations should consider investing in building a small library of books such as these that engineers can turn to over and over again, but be careful of investing in books that are unlikely to age well. The majority of books and magazines that are published, especially in printed format or archived on websites, can become dated quickly, especially if they focus on a specific version of software or technology. If you decide to invest in technology-specific books, look for electronic copies of the content as it is more likely to be kept up to date. Outdated materials increase the risk of engineers developing flawed products referencing old and possibly insecure examples.

Many of these authors have produced whitepapers and articles that cover their specialist subjects in greater depth. These sources often supplement their books with latest developments in technology and security or cover topics that were unavailable at the time the books were written. You should encourage DevOps engineers who show an interest in these industry experts to find whitepapers and other materials written by the authors and build an online library of their work.

Despite the limitation of print, it is an effective medium for those who enjoy reading. It may not suit everyone, but an accessible library containing physical copies of books that cover the topics of interest to your team members is a valuable asset. It is an effective and low-cost means of introducing established principles and ideas into a team. Factoring reading and experimentation time into daily work-streams affords team members the opportunities to study techniques which they can eventually apply to their arsenal of skills. This will elevate the quality and efficiency of their work, resulting in delighted customers.

Webinars

Live webinars are normally scheduled online events organised by established community groups or vendors. Their main objective is to offer an independent expert in a particular field, or a representative of a technology company, to present some research or new feature that may be useful to the attendees. Webinars are often free and normally take place outside of local office hours. The main appeal for attending webinars is to learn something new from people who have already faced significant technical challenges and are prepared to share their findings with their peers. There are many webinars that cover DevOps and security, and these sessions are often recorded and made available online soon after the event has finished. Adding these to your online library of learning resources is an excellent idea. During the live sessions, attendees may have the opportunity to ask questions to clarify their understanding or challenge the speaker on a particular point. Recorded sessions obviously do not allow this; for this reason, your organisation should encourage your engineers to attend live events even if they take place during the working day (perhaps because they are hosted in a different time zone). Keeping an online calendar of future and past events categorised by topic allows your team members to access resources easily. Not all webinars are well presented, so engineers who attend live events should be encouraged to write reviews so that they can be ranked accordingly.

Final thoughts on self-paced learning

Self-paced learning is better suited to the flow of DevOps practices than instructor-led training. It gives engineers constant access to the educational material needed to overcome knowledge gaps that can create bottlenecks within the delivery pipeline. It provides an environment for continuous learning, giving engineers the tools to make the right decisions to develop your organisation's products. Self-paced learning needs less financial investment than instructor-led training, even though investment in time may be higher; yet, the returns on this are huge as engineers become better equipped to design and develop secure applications and services as well as the infrastructure on which they are hosted. As with the teacher-and-student format, self-paced learning is not suitable for everyone. Some people are less proactive than others in improving their skills. There are alternative methodologies such as pair programming and knowledge sharing sessions, which I explore next.

Pair programming and peer reviews

Knowledge within an organisation can often be more extensive than first imagined. Experienced engineers within a team share a wealth of knowledge between them. They may have encountered similar challenges in previous roles, or their passion for the technologies they use promotes a disciplined approach to gaining more in-depth knowledge about them. Yet, it is common for engineers to work in isolation and never have the opportunity to impart their knowledge on their less experienced colleagues. In his book *Extreme Programming Explained* Kent Beck promotes the use of pair programming as a more productive way to deliver value through sharing knowledge and creative ideas. Extreme programming (XP) is a radical Agile methodology involving some of the purest forms of the Agile framework, while pair programming involves two engineers sitting side-by-side, using the same workstation to develop a feature together. Beck stresses that pair programming is not a tutoring exercise; however, a less experienced engineer will learn from a more experienced engineer over a period of months until the gap in knowledge between the pair is reduced significantly. In addition, pair programming improves quality because the bad habits of one engineer will be spotted by the other and corrected. Despite the perceived inefficiency of having two engineers working together, it improves knowledge sharing and reduces the risk of introducing poorer-quality code, such as code with security vulnerabilities. The two engineers can validate good coding principles, share ideas about the most secure way to configure or write code, and explore options for writing unit tests to mitigate against potential vulnerabilities.

A less extreme form of sharing knowledge between senior and junior engineers is to encourage regular peer reviews during the DevOps workflow. This normally involves a more senior engineer reviewing and validating the quality of code written by a less experienced engineer before the code is committed to a release branch. The senior engineer is likely to identify mistakes that the junior engineer has made and provide suggestions on how to correct them. The junior engineer gains new knowledge and skills and potentially avoids introducing bad habits into their work. The obvious pitfall with this form of learning is that it is dependent on the senior engineers having the capacity to review every code change made by junior members of the team. Source code management (SCM) systems can be configured with gateways that force a review before the code can be committed to the release branch. However, it is important that this does not become a 'tickbox exercise' where the senior engineer approves a code change without checking the quality of the code. One solution is to ensure code changes being reviewed are small. As Giray Özil pointed out (and repeated in *The DevOps Handbook*), 'Ask a programmer to review ten lines of code, he'll find ten issues. Ask him to do five hundred lines, and he'll say it looks good.' Using automated static code analysis to quickly identify potentially vulnerable code helps focus the reviewer's attention. I cover this in Chapter 6.

Peer reviews and pair programming are only effective at identifying potential security vulnerabilities if the knowledge and skill of the reviewer or the pair programmers are of a high enough standard. Therefore, using application security (appsec) or infosec engineers to assist in this process adds value to the process.

Appsec engineer involvement in code reviews

If you are fortunate enough to have teams of appsec engineers within your organisation, you can create an environment in which the DevOps engineers and security engineers work together to review the code-base. This could involve an appsec engineer peer-reviewing finished code or working alongside one or more DevOps engineers as code is being written. This allows appsec engineers to share their knowledge with software or infrastructure engineers to improve the security of the application and infrastructure code. For example, if a software engineer is not familiar with SQL injection flaws and has introduced this vulnerability into the code, it would be beneficial for the appsec engineer to walk the developer through an SQL injection attack scenario and explain why code written this way is poor. It is only by understanding and observing the potential dangers of writing such code, rather than just seeing the code fixed by the appsec engineer, that the developer will learn to appreciate the importance of their actions.

We should not overlook the benefits of this model for appsec engineers as well. In application development, technologies are changing constantly. Involving appsec engineers in pair programming and peer reviews gives them an opportunity to learn about the technologies and frameworks used within the DevOps value streams and better equips them to support engineers when advising them to write secure code.

As mentioned previously, the number of appsec engineers is vastly under-represented in organisations compared to the number of software development engineers. I introduced the concept of security champions earlier in this chapter. As you nurture security champions, appsec engineers can focus on educating these individuals to perform security code reviews and help identify security weaknesses during pair programming. It then becomes the responsibility of the security champion to propagate security knowledge within the team. If possible, your teams' pair programming and code review strategy will match those who have great security knowledge with those who have little or no security knowledge.

Security champions and other individuals who have learned new security skills may wish to share what they have learned with their peers using other informal forums. Your teams will benefit from having the time, space and resources to run informal knowledge sharing sessions such as regular lunch-bag meetings and short training sessions. You should also invite members of the central cybersecurity team to share some security knowledge during these informal sessions.

Informal security knowledge sharing

Sharing knowledge between appsec engineers or security champions and DevOps engineers does not require a formal curriculum. Indeed, making knowledge sharing sessions fun will encourage more people to engage either as educators or as students – or even as both. Ideally, these sessions should be informal, set up by engineers for the benefit of their peers. You should assist in making these events happen by sourcing budget for materials as well as food and drinks, and allocating time for engineers to participate, or even providing support such as booking rooms or putting up posters. The sessions should be regular, can be tied in with sprint cycles and should be open to anyone who wishes to attend, to encourage knowledge sharing across the whole organisation and not just specific engineering silos.

The form of these sessions can vary. Lunch-bag sessions obviously take place at lunchtime; they are informal sessions that usually cover a specific topic, without a formal agenda. Topics could be aligned to common security issues that the organisation's security subject matter experts can demonstrate, using the code the engineers have worked on, while encouraging the group to work on suggestions to improve the security of the code. The outcome is to enlighten individuals on avoiding or mitigating the more common vulnerabilities seen within the organisation.

Engineers should be encouraged to set up short training sessions or presentations on a specific security topic that is relevant for the team and will improve outcomes within the value stream. For example, sessions could cover implementing an application security testing tool, implications of a newly discovered CVE

(Common Vulnerabilities and Exposures) threat or describing new security features of a technology used by the engineers. These sessions allow individuals to share new knowledge, creating an outlet for new skills or technology they've learned. Seeing their peers presenting to them can also motivate other team members to follow suit and present at future events. Teaching a subject is also a great way to learn the subject.

Show-and-tell sessions allow DevOps engineers to showcase features, capabilities or methodologies they have worked on in the most recent delivery. They offer teams a chance to celebrate success with other teams within the organisation; however, the tendency within these sessions is to focus on customer value rather than non-functional requirements such as performance and security. Security should be celebrated too, and you should encourage individuals within the teams to incorporate the security measures that they have recently implemented into their presentations. Teams can, for example, share how they reduced the number of vulnerabilities or false positives reported in a scan result or how they implemented a certificate factory in a containerised environment. Sharing security skills and knowledge between value streams propagates good security practices across your organisation and improves outcomes for all your organisation's products.

Through various informal knowledge sharing sessions, we can celebrate the success of everyone involved in improving security by giving them an outlet to show off their capabilities. In turn, their peers are fully engaged in the process so that security education is spread throughout the organisation at a reasonably low cost.

Experimentation

The most effective way of learning new skills is to be hands-on. There are times when an engineer will only understand the nuances of a new technology or process through trial and error. In DevOps, value streams use experimentation to deliver the best outcomes for customers. This means DevOps teams can work on several new features or products and deliver them to small groups of customers to see how well they perform. The ones that are better received by their customers become standard features, while those that were less successful are either confined to the rubbish bin, archived or re-worked and delivered as a new experiment.

Experimentation can also be applied to the process of assessing new tools or technologies to be integrated into the product value streams. Those that make design, development and delivery of features more efficient and secure will be adopted and expanded within the team, while those that are less effective are quickly removed. For example, engineers may choose to adopt static application security testing tools (more on this in a later chapter); however, their effectiveness must be evaluated against outcomes. The ones that provide the greatest benefit in terms of accuracy (identifying genuine vulnerabilities), ease of integration (used within the IDE or build pipeline) and ease of use (identifying and fixing vulnerabilities) are adopted, while others are rejected. This approach reduces cost (because only effective tools are adopted) and improves security (everyone benefits from using the tool). You should resist the temptation to allow security teams to unilaterally assess and purchase security tools and instead create a culture in which security engineers collaborate with DevOps engineers to choose the right tools for their value streams.

Security features can also be assessed in a similar way. To provide a basic example, if engineers are undecided on which authentication methodology creates the better customer experience, they can develop two authentication features and deploy them in an a/b experiment to see which one produces the best outcome. Metrics gathered by management information (MI) tools may identify that one feature has a higher drop-out rate or takes each customer longer to action on average. These measures can be assessed alongside the benefits from a security perspective to provide a more balanced risk assessment of the feature based on business impact. In the example scenario, the impact due to the lack of customer engagement may outweigh the impact of using a feature that has reduced security but is easier for customers to use. Ultimately, understanding risks is part of the education process and this type of experimentation creates that learning experience.

Experimentation is more likely to be effective if the whole delivery team is involved. If it is assigned to one or two senior engineers within a team, it will be difficult to obtain full buy-in from the other members. On the other hand, if everyone is involved they will feel fully engaged and part of the decision-making process, as well as benefiting from learning a new skill or technology. It will also make adoption of the technology or product more acceptable to the whole team.

Learn from incidents

Incidents affecting production systems – especially security incidents – are generally bad news for everyone involved. DevOps cultivates the practice of failing fast and learning from production metrics, meaning that engineers observe the issue through monitoring and alerting, fix it and learn from it so that it is unlikely to happen again. To reap the rewards of a learning from failures, it is important to put in place the mechanisms to assert a condition and then fail when that condition is not met. For example, unit tests can be written so that the same issue can be captured during development, where feedback loops are faster and the issue is cheaper to fix.

Monitoring incidents also provides opportunities to see which weaknesses are more prevalent and suggest a potential gap within the collective knowledge of the delivery team. As a result, cybersecurity teams should be engaged to provide some training for the teams to reduce the frequency of the issue appearing in production.

Incidents are not necessarily disruptive, but they can provide insight into how well an application is performing in production if targeted by an adversary. For example, a user may have genuinely forgotten their login details and tries to log on multiple times using different passwords. This is a legitimate scenario; however, it could also indicate a brute-force attack is underway. A simple solution to this is to limit the number of login attempts for a single user, which is normally done within the application's configuration settings. But how many attempts should you allow? You do not want so few that the measure stops a valid customer from logging in, and you want to avoid a value that allows an attacker to successfully authenticate through brute force. Engineers can learn the best authentication approach through experimentation. Although this is a simple example, it highlights that monitoring security events and learning from them can improve the security of the product and produce better outcomes for the customer.

Creating a fail-fast-and-learn mechanism allows DevOps engineers to identify issues quickly and fix problems before they become bigger. They will also learn to challenge assumptions about how security features work within a system and use exception handling to assert when a security incident or a production issue affecting a security feature occurs.

Practise security incidents

It may be a frightening concept to deliberately create a security incident in production to see how resilient the running services are. The result of the incident offers an opportunity to learn how to improve resilience and thereby strengthens the services your organisation provides to its customers. This methodology was introduced by Netflix and published in a blog written by Yury Izrailevsky and Ariel Tseitlin in 2011. Netflix developed a tool called Chaos Monkey that randomly takes down a production instance to see whether the service is resilient enough to prevent customer disruption. The idea is based on the concept of a monkey in a data centre randomly pulling out cables during the business day. If Chaos Monkey does not affect the customer, the services are resilient; if there are consequences, engineers can learn from the experience and build more resilience into the service. Since the creation of Chaos Monkey, Netflix has created similar tools targeting specific abnormal conditions to test the organisation's ability to survive them. These are collectively known as the Simian Army. From a security perspective, some Simian Army tools are used to determine resilience against attacks on confidentiality, integrity and availability. For example, Doctor Monkey reduces the risk of denial of service attacks by allowing engineers to build in resilience against events that impact resources such as central processing unit (CPU) load, and Security Monkey checks for security violations, including poorly configured security groups or invalid certificates. When instances are taken down, albeit in a controlled manner, engineers learn how to address the issues, which improves their security understanding (for example, out-of-date TLS certificates are unacceptable) and teaches them how particular security incidents affect their services.

Introducing similar practices into your organisation could improve the security of applications, services and infrastructure as well as educate engineers on building applications and services with a security-first mindset. It encourages openness between security, developers and operations so everyone fully understands the impact a change in software can have on the operational side of the business.

'Play' hacking

Some engineers find that hacking an application is educational, rewarding and exciting. Obviously, I am not suggesting that you encourage members of your organisation to start hacking random websites – that will cause you a lot of legal problems; however, there are many tools available that support education through hacking. Be aware, though, that most are specialist tools used by penetration testers, bounty hunters and ethical (as well as unethical) hackers. For example, Metasploit is a commercial product maintained by Rapid7; it contains a set of tools used to break into remote systems and help companies identify vulnerabilities in their products. To support these tools, a number of deliberately vulnerable VMs (virtual machines) have been created, such as Metasploitable, that allow penetration testers or interested engineers within your organisation to hone their skills. Alternatively, your engineers can target VMs that host older, unpatched operating systems such as Windows XP.

It is unreasonable to expect software and operations engineers to be proficient in the use of products designed to educate penetration testers. However, the Open Web Application Security Project (OWASP) has, among its flagship projects, an open source web product called Juice Shop which is an application designed to help DevOps engineers learn about security. The Juice Shop application contains a large number of security flaws that have been deliberately included in the website. Engineers work through hacking challenges that allow them to exploit the vulnerabilities and, in doing so, learn how to build secure applications. OWASP also supports a lab project called WebGOAT which is another intentionally insecure application that offers users a set of tasks aimed at exploiting underlying vulnerabilities. Both tools educate developers on how to write secure software by showing them how attackers can use vulnerabilities within an application to compromise a network. I would encourage you to make these applications available to your engineers and integrate them into a gamification strategy. They are a fun way to learn how to become better engineers.

Certification

Receiving awards for achieving specific goals is a healthy motivator for many people. Runners completing their first marathon will wear their medal with pride, no matter their finishing time or position; the evidence of their achievement gives them great satisfaction. Within the professional environment, this evidence of achievement manifests itself in different ways: a bonus payment, a voucher for a family day out or a meal, tickets for a show or even a decorated lanyard ribbon can have the same effect. From an education perspective, beyond the accolade of completing a difficult course or learning a new skill, students like to receive a certificate when passing an exam or completing their study. Many tech giants such as Microsoft and Amazon, as well as specialist training and certification associations such as (ISC)², ISACA and CompTIA, offer industry-recognised certificates via their own training programmes. Some of these programmes involve considerable time investment to learn the skills to pass the relevant exams, whereas others evaluate current knowledge and skills and issue certificates to professionals who demonstrate they have the necessary qualifications to pass the exams. Despite the requirement to spend time studying for exams, the knowledge gained by students is likely to benefit their teams and the products they are working on. It reassures their customers that accredited professionals built the products and services they are delivering to them, and therefore, investing time and money for employees to gain qualifications is worthwhile.

Avoiding entropy

To remain competitive, products need to be constantly updated with new features, modified customer journeys or bug fixes. The alternative is poorer-quality products, diminishing market share and shrinking profits. Investment in products is essential to maintain a healthy market presence in which products can thrive commercially. Knowledge suffers the same fate: failure to invest in education will result in diminishing skills, poorer products and a disenfranchised workforce that will eventually harm profitability. If you invest in education, you will maintain the knowledge and skills to deliver better products, allowing your organisation to maintain its competitive edge.

Entropy describes loss of information – in this case, knowledge and skills. It is caused by two factors: *attrition*, when staff leave the organisation, taking their skills with them; and *technological advancement*, when technologies and processes used for developing applications and software progress faster than those used within an organisation. The two components of entropy are intertwined; as engineers find their skills are not keeping up with technological advancements, they are likely to leave the organisation to seek opportunities to improve their skills elsewhere.

Continuous investment in education is necessary to mitigate against entropy in your organisation. This is not a one-off exercise, or a regular catch-up programme: education is an ongoing process. It needs to be embedded in the daily activities of the organisation. From a DevSecOps perspective, engineers need to engage in a continuous programme of improvement to make use of the latest technologies and processes to deliver high-quality and secure products that will continue to delight your customers. DevOps engineers need to continuously research the most effective development practices using the most suitable and reliable education programme. Since individuals respond differently to the various ways of learning described above, it is advisable to adapt multiple approaches and review them constantly with staff members. The goal is to constantly improve the knowledge and skills of the organisation as a whole. The lower the *entropy* of knowledge and skill, the more likely your organisation will deliver high-quality, feature-rich and secure products.

Measuring entropy is difficult, although there are some techniques that can be used. Let's imagine, firstly, that the number of injection attack vulnerabilities in your organisation's production code is too high. Secondly, let's say the level of knowledge of each individual in the team who can design, develop and test controls that mitigate this threat is known. Finally, let's say you've calculated the probability of these members leaving the team based on the team's attrition rates.

Over time it is evident that, as the probability of individuals leaving the team increases, the knowledge and skill within the team required to fix this defect will also decrease. As this skill value tends to zero, the greater the risk to your organisation. Knowledge is not lost exclusively through members leaving the team; it can also diminish over time as team members forget what they have learned previously. If your organisation does not invest in continuous learning over time, it will lose the knowledge and skills needed to maintain up-to-date and secure applications and infrastructure, resulting in lower customer satisfaction and reduced market share. On the other hand, if you maintain the level of knowledge within the team, the risks remain small.

Conclusion

Education is at the foundation of DevSecOps: it is explicitly identified as one of the three ways of DevOps. Everything DevOps engineers do is a learning experience, whether it involves learning new technologies or processes, learning about their customers' behaviours and how changes to the applications they develop affect them, or learning how to develop secure systems. Cybersecurity is everyone's responsibility, not just that of a central security function. For DevOps to be effective, everyone in the value stream must understand their role and their responsibility to develop secure products. They must have an appreciation of how threat actors exploit vulnerable systems so that they can defend against them.

In this chapter, I have identified a number of methodologies to integrate security knowledge into the value stream. I do not suggest that everyone working in DevOps be experts in security; instead, I proposed that the collective knowledge within a DevOps delivery team is comprehensive enough to deliver secure software. I suggested that those who show a willingness and capability to learn over and above the security requirements of their role are prime candidates to become security champions and propagate knowledge throughout the team. Finally, I discussed the perils of letting entropy reduce the security capability within the value streams.

In the next chapter, I move on to the next layer of DevSecOps, secure by design, which builds on the knowledge acquired through a robust security education programme.

FIVE

Layer 2: Secure By Design

As we explored in the last chapter, laying a foundation based on good-quality education is crucial for the adoption of DevSecOps practices. The provision of relevant resources that enable engineers to learn their trade and take responsibility for security in their day-to-day activity is a necessary investment to continue with DevOps practices. Empowering engineers with security skills and knowledge at the local level supports the flow and frees DevOps teams from bottlenecks caused by security gates within the pipeline.

On top of this education layer, I propose a second layer in which engineers must apply best developer practices in their work to design security into the applications and services they develop as well as the infrastructure on which these products are hosted. This is a large topic in its own right and there are plenty of resources available to help developers write quality code; therefore, I will focus on some of the key features that DevOps engineers should focus on in designing quality systems.

Introducing good design principles is built upon a strong educational foundation. Without the knowledge and skills to architect and develop secure software, it is less likely that engineers will apply security at the core of their development. Therefore, jumping to this layer without building an education programme first will be ineffective. Even if engineers are working on a new concept, it is essential that they have built their understanding of the security requirements in relation to the concept and addressed any shortcomings in their collective skill sets.

The importance of good design principles

Integrating good design principles into the architecture and development of a product is a fundamental requirement for writing secure code. Application code and infrastructure code, when written well, are a line of defence against the multiple threats from malicious actors. Conversely, implementing poorly designed applications, services and infrastructure can be your organisation's Achilles heel, exposing your company and your customers to cyber-attacks. Your engineers should have an understanding of the basic principles of good coding practices and know what poor coding practices look like in order to avoid them in their own work.

Security defects are like any other defect. Some are obscure, caused by nuances of the underlying technology being used within your organisation. For example, if the coding language is case insensitive, mistyped parameter names can cause bugs to appear in the application or service. Many other defects are introduced by low-quality architectural design or coding errors that cause the application to fall into an unexpected or unstable state, which is then exploited by hackers. A common example of this type of defect is a buffer overflow that allows hackers to take control of the memory stack of an application. Controlling application state at all times during its execution is critical. This involves explicitly determining the outcome of each decision point within the application flow. It is common for engineers to focus on the happy path and not consider bad paths caused by invalid user input, unavailable resources (such as APIs and databases) or unexpected exceptions.

Security design extends beyond the application to containers and infrastructure. Good design principles are essential in building a container-based architecture and its underlying hosting platform. Default settings in many container technologies are not secure, which means that consideration must be given to making informed decisions about the way the system is designed and implemented.

Making sure that engineers adhere to good design principles carries greater importance when the architecture relates directly to security controls. If the security controls are weak, they are easy targets for hackers.

These controls are prevalent throughout the value stream, from those built into the application during development to those that protect the application at runtime. These include securing the source code or securing the application runtime. They also extend to customer interactions with applications, such as authentication and authorisation journeys.

A robust design is based on simplicity. Over-engineering a problem leads to a lot of complexity when writing software. I often observe developers using clever workarounds to solve technical problems rather than focusing on the simpler solution to meet a business requirement. A similar scenario occurs in infrastructure as operations staff build complex rules when configuring an environment. Unfortunately, these nuanced solutions are rarely documented, which means that other members of the team or people in different teams have a hard time trying to understand how a piece of code is written or how a network is configured. It is well known that increasing complexity increases the risk of introducing vulnerabilities that are difficult to find (although simple for hackers to exploit).

The use of open source software is a common solution to reduce complexity, although, as we shall see in Chapter 6 when discussing software composition analysis, this introduces another level of risk if not managed carefully. Software engineers often search public repositories to locate an existing third-party solution that solves a particular problem the engineer faces. Incorporating these open source components into the codebase reduces the complexity of the task the engineer is trying to emulate.

Engineers should also have their code peer reviewed (as discussed within layer 1) to check that there is no undue complexity. During the review, engineers should be able to explain the business reason for the way the code is written. If they spend excessive time explaining how the code is 'plumbed in', they have likely spent too much time over-engineering their code.

I have come across 'securely designed' architectures that have been implemented with a delivery-first focus, where security takes a back seat. This is common when engineers are pressured into releasing new features or products to market with tight time constraints. The result is an increase in technical debt, which leads to a downward spiral of poorer code quality and system security.

In summary, building on the education layer to create a layer of robust security design principles adds further strength to DevSecOps as a practice.

Threat modelling

Although I have separated education and security design principles into two distinct DevSecOps layers, the glue that joins the two together is threat modelling. Threat modelling is at once an exercise in education and a process of making the right security design decisions.

DevOps engineers should be familiar with the types

of security weaknesses that attackers search for so that they can mitigate against them being exploited. When designing a new product or a feature within an existing product, it is essential to understand the risks inherent within the design so that they can be eliminated or mitigated early in the delivery lifecycle. Threat modelling is a process in which engineers assess the design of the product or feature to identify threats and to work out how to build in protection against them. There are several documented methodologies, including one described by Adam Shostack in *Threat Modeling – Designing for Security*, which introduces a four-step framework designed 'to align with software development and operational deployment'. The four steps are:

1. What is being built? – Model the system being changed
2. What could go wrong? – Find threats using a recognised threat model methodology
3. What could be done about it? – Address the threats based on risk appetite
4. Is the threat analysis correct? – Validate the threats with a security testing programme

To obtain the best results from threat modelling, all DevOps engineers working on the product or feature to be assessed should dedicate a significant period of time to the exercise, preferably co-located in a room with a whiteboard. The process starts by having engineers draw a data flow diagram (DFD) of the product, paying particular attention to where data is at rest or in transit, and mapping the security boundaries across which data flows, such API contracts, as well as the actors who should have access within these trust boundaries. Focusing on data is important because an attacker's goal is to target data; for example, exfiltrating personal details (including credentials) or manipulating data such as shopping baskets, financial transactions and even source code.

Trust boundaries define areas that share the same level of privilege and where changes in privilege occur. These are important to identify because attackers look for ways to access different areas of a system using weaknesses in privileged access control. For example, customers have access to view their own shopping basket but do not have access to other customers' shopping baskets. However, the fulfilment team needs access to view many customer shopping baskets in order to process their orders. Attackers search for weaknesses that allow them to assume the identity of a customer but gain the elevated access rights of the fulfilment team to view all customer shopping baskets. Data should be protected behind different trust boundaries implemented by robust access control policies. DevOps often involves small incremental changes to an application; if there are no updates to the flow of data in and out of the security boundaries, or there are no amendments to existing security controls, the threat modelling exercise can focus purely on assessing the change in the context of existing architecture.

Finding threats is a complex process which, if not carried out correctly, can lead to missed potential weaknesses in the application or service. As discussed previously, there are three types of attacks: those on confidentiality, those on integrity and those on availability. The main objective of threat modelling is to identify how the application may be exposed to these types of attacks. In addition to the CIA triad, there are also elements of trust that require threat modelling to identify weaknesses associated with authentication and authorisation. There are several threat modelling techniques, such as STRIDE, LINDDUN, PASTA and VAST, each with its own set of advantages and disadvantages. The most mature methodology is STRIDE, which has been around for two decades at the time of writing this book.

STRIDE

STRIDE was invented in 1999 by Loren Kohnfelder and Praerit Garg. It is an acronym in which each letter describes a particular type of threat:

Spoofing threats are those in which an attacker assumes the identity of another person or another process or machine. Spoofing is a violation of the authentication property.

Tampering threats occur when an attacker modifies the content of a file, alters items stored in memory or reconfigures network settings. Maintaining integrity mitigates against this form of threat.

Repudiation threats involve a person claiming to not be responsible for an activity they carried out. Non-repudiation normally involves logging events and using digital certificates with the purpose of enforcing trust.

Information disclosure threats relate to attackers having access to data which they are not authorised to see. Mitigating against information disclosure requires maintaining confidentiality.

Denial of service (DoS) attacks lower the capacity of resources such as memory, CPU and network, reducing the ability to provide a normal service. DoS is an attack on availability.

Elevation of privilege involves authorising someone to do something they are not supposed to do. Elevation of privilege violates the authorisation property.

Discovering threats at an early stage of the development lifecycle is crucial to keep costs down and equip engineers with the knowledge to build a product or feature securely. Early feedback loops are a fundamental requirement of DevOps.

Addressing threats does not necessarily involve making changes to an application or service that has been assessed. Indeed, there are several ways to manage risks associated with the threats identified

S POOFING

T AMPERING

R EPUDIATION

I nFORMATION DISCLOSURE

D ENIAL OF SERVICE

E LEVATION OF PRIVILEGE

STRIDE

in the previous exercise. One option that is not viable, although it is common, is to ignore the risk – but organisations ignore risks at their own peril. It can lead to regulators issuing punitive measures, litigation from stakeholders affected by the risk, or even damage caused by cyber criminals exploiting the open risks. Mitigating risks involves either removing the feature or item that has associated risks or adding controls (either within the feature or within processes involving the feature) to remove or reduce the risk to an acceptable level. Mitigating the risk can be more costly than if the risk is exploited, especially if the likelihood of an exploit is significantly low or its impact is minimal. In this scenario, the onus should be on the key stakeholders to decide on whether the risk is acceptable to the business. This is not the same as ignoring the risk. Risk acceptance should be clearly documented and re-evaluated on a regular basis to establish whether any change in the likelihood or impact of the risk has changed over time. Occasionally, the risk can be transferred to the end user, which may involve setting out terms and conditions for using the affected product or service.

Testing threats that have been mitigated should form part of the strategic testing programme. Automating tests to evaluate the controls that have been put in place to mitigate risks provides DevOps engineers instant feedback in the event that a subsequent change to the application or infrastructure impacts the mitigating control. As I will demonstrate in Chapter 6, not all tests can be automated, so it is important to use manual penetration testing to determine whether the risk can be exploited. An identified threat may not be exploitable due to other controls that already mitigate the risk of an exploit. Conversely, a risk that is considered low may be combined with other low-rated risks, which collectively raises the threat to a level that is less acceptable to the business stakeholders.

Threat modelling is a useful DevOps tool in the ongoing battle against hackers. It helps to identify potential exploits quickly so that they can be managed accordingly. By bringing engineers together to discuss threats, they share knowledge of the system and gain a better understanding and awareness of security.

Clean code

In his book, *Clean Code*, Robert C Martin defines a number of principles for writing code that is well structured, concise, and easy to read and understand. Martin does not explicitly cite security as the main reason for writing clean code, but poorly written code is a significant factor negatively affecting business outcomes, including an increased risk of security incidents. The adoption of clean code practices within the DevOps engineering team is a simple yet effective way to reduce security risks within your organisation's applications.

Substandard code is synonymous with technical debt. Unfortunately, maintaining code of inferior quality introduces even more technical debt. Ultimately, this will cripple a software engineering organisation's ability to deliver high-quality features and products to an expectant market. Clean code is an enabler of the five ideals of DevOps discussed in Chapter 1: it introduces simplicity, allows developers to focus on outcomes, creates a platform for minimising technical debt (maximising psychological safety) and ultimately allows engineers to satisfy the needs of their immediate consumers.

Martin describes a number of principles, patterns and practices within his book on writing clean code. Each plays a part in mitigating against writing vulnerable software as well as infrastructure code. Not all teams have the luxury of developing new products from scratch, which we call greenfield products;

therefore, if your DevOps engineers are working with poorly written legacy code, I would advise them to refactor the code to make it cleaner. Although this effort may sound like a waste of investment, the long-term returns will include fewer security defects and faster software development.

Naming conventions and formatting

Readability helps engineers understand what the code does, so it follows that it is easier for engineers to work with code that is easy to read. Using a consistent naming convention, making use of words that identify methods and parameters, makes the code clearer for engineers; verbs describe functions which perform a specific task, while nouns describe input parameters required to perform a task. Conversely, using abstract and inconsistent naming conventions will lead to confusion that may inadvertently expose confidential data or affect data integrity. Using words that are relevant to the business domain makes it easier to see how a piece of code maps back to customer journeys, making changes to business rules clearer in the code and less prone to errors. Self-documenting code reduces the need for in-line comments. This avoids situations in which poor naming conventions and misleading and outdated comments further confuse developers.

Code is easier to read and edit when it is formatted correctly. Although most IDEs automatically assist the engineer in formatting the code, such as by indenting hierarchical statements, colour coding and using other visual aids, engineers must be wary of writing code that is poorly structured.

When code is more difficult to read, it is more difficult to maintain, and mistakes will create opportunities for hackers. An engineer may not appreciate that a parameter named 'd' is a date value and ends up allocating a non-date value, which an attacker may use to cause an unexpected error; or validation of the value's format is not performed correctly, resulting in the risk of an injection attack.

Concise coding

Whenever engineers write code, they need to make it concise. A function should not perform more than one task, and a function should not be repeated (Martin calls this 'DRY': don't repeat yourself). The more complexity an engineer adds, the more difficult it is to maintain a known and controlled state. It is common to see two functions or more with identical intentions generate different results because only one was edited in response to a requirement change. Confusion within the source code creates unexpected runtime behaviours that hackers can exploit.

Data abstraction

Hiding the database structure is a key factor when writing secure code. By exposing the database schema through public interfaces, you are giving hackers knowledge about what data is stored and their relationship with other data objects. It can also lead to confusing code in which data objects representing data structures are chained together in order to provide a representative model of a complete object. The following simple example provides a lot of clues about the structure of the database and other objects:

```
<CustomerAddress = getCustomer(customerid).
getOrder(orderid).getDeliveryAddress>
```

Compare that to the following line of code:

```
<CustomerAddress =
getCustomerAddress(Customer)>
```

In the second version, there are fewer indications of how the data is structured.

Boundaries

Using third-party libraries in software has become a necessity in recent years to let engineers focus on the features that add value to the product while leaving grunt work to software written by open source communities. We discuss open source libraries from a pure security perspective later in Chapter 6; however, in the context of clean code, it is preferable to avoid direct communications at the boundary level between your organisation's code and that of the third party. If engineers are given free rein to access the interface of the open source library directly, they are susceptible to changes made by the community developers affecting the functionality of your applications and services. For example, if a method signature changes, anywhere the method is called will have to be changed too, otherwise an exception may be thrown or an unexpected result is returned, leaving your application open to potential vulnerabilities. Martin suggests writing an abstraction method for the third-party library, meaning the code only needs to be changed in one place. Writing tests to validate this abstraction provides a safety net when changes occur, allowing DevOps engineers to respond to these changes quickly.

Concurrency and thread safety

Executing code asynchronously is a technique used to improve performance. It means that instead of function B waiting for function A to execute before it runs, functions A and B run at the same time. If the application is dependent on the execution of code in a specific sequence, your engineers need to write synchronous code, and thread safety allows shared data to be altered by one function without interference from the other when running asynchronously. However, thread safety is not easy to implement; it often leads to unexpected behaviour within the running application or service. Race conditions are typical when multiple threads make changes to the same parameter. Hackers can exploit race conditions for their own benefit, so it is important your engineers have an understanding of concurrency in the languages and frameworks they use.

Common weakness lists

Lists of common weaknesses are useful tools in the fight against cybersecurity crime. They help engineers identify the most common mistakes made in application development that are regularly exploited by attackers. They not only show you what each weakness looks like, they also provide practical advice on how to avoid introducing these weaknesses into the code in the first place. I like to put these up as posters in places where engineers gather, such as at the water cooler station or next to Kanban boards, where they are highly visible. Armed with a list of these common weaknesses, engineers can design and test applications to eliminate them from the applications they develop. The two most widely used publicly available sources are the OWASP Top 10 lists and the MITRE Top 25.

OWASP Top Ten lists

The Open Web Application Security Project (OWASP) is a non-profit foundation set up to help organisations improve the security of their software. It runs a number of community-led projects categorised by maturity, including maintaining top-ten lists of the most critical security risks affecting organisations. At the time of writing, their flagship projects include a list of the top ten mobile application threats and the highly regarded Web Application Top Ten. The OWASP lab projects have a list of top ten API threats listed under their API Security Project, while their incubator projects have a number of top-ten lists targeting Docker and Serverless technology. These lists are updated approximately every three or four years to reflect the latest trends in application risks. The flagship OWASP Top Ten project is considered by many to be the essential list of the most relevant weaknesses affecting web applications.

The lists are compiled from surveys in which the respondents are asked to rank the top four security weaknesses that they believe should appear in the top-ten list. Many organisations across multiple industries and geographies participate in the survey. As a result, the security industry considers the OWASP Top Ten list representative of the most common security threats affecting businesses today.

The OWASP Top Ten list (version 2017 at the time of writing) is freely available as a pdf document. Each of the top ten weaknesses is presented on a single page in a consistent format. The first part rates the general risk in terms of exploitability, prevalence, detectability and technical impact. This rating is not in context of specific applications or businesses, but it helps you assess the risk rating in relation to your organisation. The rating is then followed by four summary sections that describe which applications are vulnerable, examples of an attack scenario, preventative measures, and references to materials relating to the risk. It is important for

DevOps engineers to be familiar with each of the items on the Top Ten list and factor security into their design.

The OWASP Top Ten list of web application security risks is arguably the most important project for OWASP. It has spawned many other projects to help engineers design security into applications they are developing; for example, OWASP recommends you use the Application Security Verification Standard (ASVS) as a guide to designing secure applications. Their projects have also produced a series of simple good-practice guides called OWASP Prevention Cheat Sheets, which show engineers how to design and implement security from the beginning, and the OWASP Top Ten Proactive Controls, which is a list of security techniques that all engineers should follow to build security into the products they are developing.

All of these projects are available for free on the OWASP website, and you should encourage all your DevOps engineers to use these tools to become better equipped to build security into their products. The OWASP tools help engineers overcome limitations within the languages and frameworks they use, which are often designed to make development easier but not necessarily more secure.

OWASP security knowledge framework

Among OWASP's flagship projects is an open source education tool that teaches engineers how to prevent hackers from exploiting security defects in your organisation's products and services. The OWASP security knowledge framework (SKF) uses the OWASP ASVS to guide engineers through building secure software. This tool can be integrated into the workflow to enable engineers to securely design and develop software, while offering them security labs to enhance their security skills.

The SKF project consists of a dashboard that can be downloaded from the OWASP website. Once installed, engineers can set up projects and users within the framework which are aligned to a customisable list of categories, such as web applications or mobile applications. The framework contains a series of checklists for various requirements within the ASVS that can be assigned to your categories along with detailed knowledgebase articles explaining how to tackle the likely attack vectors associated with each type of exploit. Engineers also have access to code examples for a number of different languages plus a lab in which they can practice some of the techniques used by hackers and how to defend against these types of attacks.

MITRE Top 25 Common Weaknesses

In addition to the OWASP Top Ten lists, MITRE, in collaboration with the SANS Institute, maintains a list of the twenty-five most dangerous software errors which contains the 'most widespread and critical weaknesses that can lead to serious vulnerabilities in software'. As with the OWASP Top Ten lists, this is a community resource with contributions credited to a large number of people in the software and security industry. The list is ranked according to the Common Weakness Scoring System (CWSS), which standardises the vulnerability rating based on three groups of metrics: *Base Finding*, covering the inherent risks, confidence of the accuracy, and strength of existing controls in relation to the weakness; *Attack Surface*, or the barriers that the attacker must overcome to exploit the weakness; and *Environmental*, which examines the operational and business context of the weakness.

The SANS Institute offers a number of resources for organisations to help reduce the occurrence of the top twenty-five programming errors. However, unlike OWASP, which supports a large number of free security-related projects covering tooling, guidelines and support groups, the majority of the resources from the SANS Institute focus on selling commercial application security training and security awareness products, including publications outlining industry standards. Nevertheless, these are useful resources that promote the practice of secure by design.

Final thoughts on common weakness lists

As would be expected, there are many overlaps between the MITRE and OWASP lists, although they have their differences. Therefore, it is advisable that engineers, product owners, testers and project managers are familiar with both sets of lists.

These lists are the product of interactions with a large community and are indicative of the most widely exploited vulnerabilities, so it makes a lot of sense to ensure all engineers are familiar with their content. Consider printing them out (the PDF version of the OWASP Top Ten Web Vulnerabilities is formatted in such a way that each vulnerability can be printed onto a single A4 sheet of paper) and placing them around the office where they will be easily noticed, such as around water coolers, coffee machines, vending machines, communal kitchens and even toilet facilities. This creates an environment in which engineers openly discuss common weaknesses including how to design applications to avoid introducing them into the products they are developing.

In support of these lists, there are application scanning tools which scan software to look for signatures that indicate the presence of these vulnerabilities. I will discuss these in Chapter 6.

Core application security design principles

There are a number of features that I consider to be at the heart of addressing the OWASP Top Ten application security risks. I believe DevOps teams should invest in learning about these security design principles and apply them within the applications and services they develop. These are different from the concepts of writing clean code discussed earlier because they incorporate cross-cutting features and controls. Factoring security design principles will help mitigate against many of the attack vectors used by hackers.

Validation

The primary reason for validation is that you must *never* trust user input. Whenever an application expects a parameter, such as from an input field on a web form or from a query string from an HTTP request, developers should always implement a mechanism to validate these inputs to protect from nefarious activity. Using domain primitives (as mentioned later in this chapter) provides developers a certain level of control to manage input values and the state of the primitive object. As we have seen, some of the OWASP Top Ten vulnerabilities involve failures to validate input data. In SQL Injection and Cross Site Scripting (XSS), an attacker could manipulate user input to cause an application to behave in a way it is not supposed to.

In a multi-tiered application, there are multiple layers each requiring its own set of defences. It is a common mistake made by developers to validate only one layer of the application. For example, a standard application has three tiers: the frontend presentation layer, the backend server layer and a database. The easiest layer to validate data is at the point of user entry. Therefore, developers tend to put all their validation into the frontend forms to check the validity of the data users enter. Unfortunately, less emphasis is put on validating the inputs at the server and database layers. The engineer makes the basic assumption that all traffic arriving at the server is validated by the front-end client. Unfortunately, this cannot be guaranteed. An attacker has access to a number of proxy tools to intercept requests to the server and change values after they have been validated by the client. This allows the hacker to bypass the client validation process and send unvalidated data to the server. Obviously, if the server is not validating this input, the attacker is able to inject non-validated values into the unprotected server. Likewise, if the database does not validate these values, the data will be inserted into the database. Why is this the wrong approach? If an attacker is able to insert any value into the database, it could be used to run scripts (such as XSS attacks) from subsequent page requests, alter data including personal or financial information, or corrupt the content of managed websites.

It is important that validation logic remains consistent across all layers of the application. If the rules between the different tiers are different, it could create defects that cause unexpected outcomes. For example, if an allow-list of valid characters is used for client validation and a block-list used for server validation, there could be conflicts in which characters are permitted. If validation rules change, it means they have to be updated in multiple places. Domain primitives provide an excellent mechanism to build validation into all layers of the application with minimal effort. The authors of the book *Secure by Design*, Dan Johnsson, Daniel Deogun and Daniel Sawano, describe how domain primitives work; I provide a brief summary below.

Domain primitives

As we have just seen, an important mechanism for protecting applications is to validate all input data across the boundaries of the application. This means checking the format, whether it comes directly from a user, from a dynamically constructed query string, or from another endpoint or API. A feature expecting a string of a certain length with a limited character set should only allow strings that meet this criterion. All other values should be rejected.

Let's take the example of a UK postcode. This value has a specific format that follows a documented standard. Most UK postcodes consist of one or two letters at the start, followed by a number from 1 to 99, followed by a space, another number from 1 to 9, and two more letters. There are no non-alphanumeric values in the format. Each time a user enters a postcode, the application needs to ensure that they use the correct format; submitting an incorrect format may indicate an attempted injection attack. The string may include the pattern of an SQL injection attack, or the user may enter a long string that takes a long time to process (which may indicate a denial of service attack). Your security-educated engineers know that it is important to validate the submitted string, so each time a postcode string value is used in the application it is validated. However, if the postcode value is used in multiple places within an application – which, in a DevOps organisation, is developed by different engineering teams – it is likely that not all instances of the postcode are validated correctly. If the UK Post Office changes the postcode format, or the product is made available in a region that uses a different format, all code that validates the string will need to be changed. Unfortunately, inconsistencies may creep in or some are not validated at all, resulting in your products accepting incorrectly formatted and potentially malicious postcode values.

A more elegant approach is to use *domain primitives*. The authors of *Secure by Design* define these as a combination of 'secure constructs and value objects that define the smallest building block of a domain'. Returning to our postcode example, instead of using a string object, the developer creates a domain-specific object to manage postcodes. This object is self-validating, and it will only be created if the format of the postcode provided by the user is correct; if the postcode format is incorrect, the object cannot exist. However, once the object has been instantiated it is passed as a parameter within the application. This ensures that the postcode value is always in the expected format. Finally, if the format of the postcode needs to change, the developer only needs to alter the validation code in one place. Designing applications and services to use domain primitives is a simple solution to a difficult problem and enhances the security of your organisation's products.

Regular expression validation

The validation of strings often requires the use of regular expressions. Within the domain primitives, you may need to validate the conversion of the string value to the domain-specific value against a regular expression. However, developers must use them with caution because they have their own security weaknesses if not used correctly. To avoid one potential issue, developers should validate the string length before parsing the string through a regular expression validation function. The reason for this is that longer strings take an exponentially longer time to perform the regular expression validation. Hackers often exploit this weakness to perform a DoS attack against the application performing the validation. Returning to our postcode example, your engineers need to check that the string is no longer than eight characters before stepping into the regular expression validation.

Access control policies

For hackers to reach a target within the organisation, they need to gain access to a number of resources as they navigate through the network. Their primary objective is to gain administrative access to servers in order to distribute malware, run their tools or exfiltrate documents, among other activities. If they gain administrative access to the network, it's 'game over' for your organisation. The responsibility of defending against this threat is shared between developers and operations. They must design systems that restrict access to the various items within the network according to their sensitivity. Access control policies define the *principal*, which is the entity requesting access (such as a user or component), the *subject*, which defines the active entity performing the request (such as a running process), and the *object*, which is the passive entity to which access has been requested.

Access control policies enforce rules that determine the identity of entities (authentication) and the permissions afforded to each entity (authorisation). These policies play an important role in the design of an application and system. If there are any weaknesses caused by insecure authentication and authorisation, threat actors will exploit them to gain control of network services of interest to them. The following subsections dig a little deeper into authentication and authorisation as well as covering the topic of accountability, a topic related to access control.

Authentication

Many applications need to know that they are interfacing with a specific known entity. That entity could be a person (such as a customer, a registered user or a member or staff) or a component (such as another API, endpoint or application). Entities should only grant access to those they trust; therefore, it is crucial to establish the authentication architecture for the entire ecosystem before it is implemented. You need to be absolutely certain of the identity of the resource, whether it is a user or a component, before granting them access to your valuable resources. If the entity is a person, authentication uses four factors to establish the identity of the user:

1. Something they know, including a password or a personal identification number (PIN)
2. Something they have, such as a mobile phone, an authentication token or an identity card
3. Something they are, determined by biometric authentication, such as a fingerprint or facial features
4. Something they do, such as keystroke behaviour, voice recognition or physical signatures

Ideally, you should use at least two factors in determining the identity of a user, preferably more for accessing the most restricted areas of a system. This is known as multi-factor authentication (MFA). There are several common MFA implementations which your DevOps teams should be encouraged to adopt in developing their applications and building the infrastructure. For example, once a user has provided a username and password (what the user knows), they receive a one-time password (OTP) that is sent to their phone (what the user has) via SMS. To complete the authentication journey, the user must enter the OTP into the web browser or application. A recent threat against this method is SIM-jacking, which involves an attacker taking ownership of the mobile telephone number, allowing them to intercept the OTP message sent to the phone. The solution is to ask the user to enter an OTP generated on a physical access token or by an authentication provider installed on their phone, which bypasses the need to use SMS messaging.

When designing a product or feature, or even a process within the value stream, you should determine whether you need to implement authentication. In some cases, the employment of anonymous users is valid; for example, if your products are informational, such as a news portal. However, there are many instances in which users of an application need to register in order to perform a number of tasks. A typical model would be an application in which customers buying a product from a website on which they have already registered need to log on in order to access their payment details and delivery address. The customer's experience of using the website is enhanced because they do not have to retype this information each time they visit the site to buy a product or service.

Not all principals are human, which means some of the authentication methods are irrelevant, such as biometric and behavioural authentication. Non-human authentication involves establishing trust between entities – a process that often uses public key certificates such as X.509. Because these certificates are public, they are not a proof of identity unless the entities know the private key corresponding to the public key. Of course, management of the private key is crucial to avoid it being exposed and exploited by hackers. Keys must never be hardcoded within the entities, even if they have been encrypted. They must be kept in a separate secure store, such as a secure password repository or a hardware security module (HSM).

Your engineers need to design applications and services with a high degree of assurance that the identity of the principal is legitimate. Further, as systems become more distributed, it is critical to establish identities and authenticate them across multiple services. This increases the complexity of the authentication process. Fortunately, there are a number of protocols in use today that simplify the task, including IdP, SAML, and OAuth supporting OpenID Connect. DevOps engineers should be familiar with these technologies and apply them accordingly.

Authorisation

Authorisation determines which objects a principal has access to and the manner in which it can access each object. The key to a strong access control policy is to apply the principle of *least privilege*, to allow only authorised access to resources required to perform an assigned task. There are several protocols for controlling access to resources, such as access control lists (ACL), role-based access control (RBAC) and attribute-based access control (ABAC). Designing access control policies is a complex activity; it is common for engineers to ignore the principle of least privilege and grant higher-privileged access rights to reduce this complexity, particularly during development and testing. This is a high-risk approach to managing authorisation that opens up opportunities for hackers to exploit, so engineers must have a comprehensive view of the identities, roles and role functions to implement a robust access control policy. This understanding applies across the entire delivery pipeline protecting all resources within the value stream, including documentation, source code, container images, configuration files, automated testing tools and deployment tools, as well as within running applications. In short, every entity that needs to interact with another entity must be governed by rules that determine the nature of the association.

Accountability

Although authentication and authorisation provide a mechanism to protect resources, accountability offers reassurance that the actions of principals can be traced and that they comply with the access control policies. Designing an ecosystem which logs authentication and authorisation journeys is essential. They offer insights into the actions of entities, including employees, customers, suppliers and running applications, to support non-repudiation, detect intruders and deter inappropriate use of access controls. Detailed access control logs are also essential if legal action is to be taken against individuals acting fraudulently.

Immutable objects

Designing security into application classes is another area requiring some attention. I have worked with developers who try to do too much with their classes, and I've been guilty of this too. We end up creating a bunch of public accessors in our classes so that consumers can alter the state of an object after the object has been instantiated. Thus, a class may have a number of public setters that allow a user to change the value of the property at runtime. For example, if developers create a *customer* class, it is considered practical to provide a writeable *name* property in case a customer needs to change their name (following a change in marital status, for example) or a writeable *address* property in case the customer wants to change their delivery address. However, public accessors provide no control on how or when these values are altered. A hacker can use a public accessor to change the value of the instance property such as the *address* value during the final stages of an order fulfilment process. Instead of the customer receiving the goods, the hacker diverts the items to another address by setting the *address* value during runtime. To counter this risk, *customer* objects that are created for use by this fulfilment process should remain immutable. Developers should design classes in such a way that the objects are pre-populated with correct values only during instantiation. These properties should also be read-only, which means there is no way to change the state of the object at runtime. As a result, hackers cannot change the state of an existing object. In the above example, the delivery address remains as the customer intended.

Error handling

It is impossible to guarantee a completely error-free application during runtime. Exceptions occur for various reasons: resources unavailable, unexpected values, null pointers, etc. Robust applications are able to manage exceptions when they occur so that the state of the application remains predictable. The application or service must also offer a positive customer experience at all times, so when an exception occurs the customer is presented with a helpful message that allows them to navigate to a safe zone. It is all too common for users to be presented with a page containing irrelevant technical details, or even a blank page, with no means to navigate away. If this occurs during a payment transaction, it can be unnerving for a customer, who will be unsure whether payment was taken or an order placed. If exceptions are not handled correctly, hackers can exploit the unexpected state of an application to control its behaviour to their own nefarious advantage. The aforementioned page that shows a technical breakdown of the exception is not useful to a customer, but it is a goldmine to a hacker. The information may give away clues about the underlying architecture or network topology, such as function names, class structure and IP addresses. The hacker can also use thrown exceptions to locate and exploit blind injection attacks. Good design principles encourage programmers to use exception management correctly and your engineers should be familiar with these principles. For example, it is imperative that all code that could throw an exception is handled within a *try-catch-finally* statement (or equivalent).

Your development engineers who write application code should decorate exceptions with relevant details to help DevOps engineers – including other developers, as well as operations engineers – to fix issues causing the exception. Generic exceptions are largely unhelpful, forcing engineers to explore the root cause of the problem through a process of debugging, slowing their ability to fix the issue. Likewise, the use of error codes should be treated with caution. Using exceptions to log error codes causes confusion and slows down debugging and defect fixing, affecting flow. As I have mentioned previously in Chapter 1, disrupting flow within the value stream could have an impact on the core business.

Using exceptions to control business decisions will also have a negative effect on productivity. This is where an exception is thrown when a particular state is reached, and the exception handler is used to run another function. For example, the following pseudo-code should be avoided:

```
<if value=true then return value else
throw new Exception>
```

This *style* of coding makes it more difficult to understand the logic, increasing the risk of failing to control the state of the application. The result is security vulnerabilities are much harder for engineers to find within the application. A better solution is to use logic in which the value is handled with logic:

```
<if value=true return value else
doSomething()>
```

Logging exceptions

When exceptions occur, it is important to log the details in a suitable logging tool. Error messages produced during an exception must contain enough information to help developers, testers or operations staff resolve the problem that caused the initial error. As stated in the 'Error handling' subsection, the customer does not need to know the low-level details of what went wrong within the application. They will benefit from a friendly error message with a brief explanation of what went wrong and offering a way out of the situation. On the other hand, engineers must have as much detail as is necessary to resolve the problems causing the exception; therefore, providing a full stack trace is useful for the engineers. However, it is worth repeating here that hackers can mine details from stack traces that give clues to the design of the application and the environment on which it is hosted, so it is imperative to add controls to defend the logs from attackers. Highly sensitive data must be encrypted or obfuscated (ie masked with characters such as '#'). Access to logs must be restricted and time limited, preferably with a privileged access management system.

Within DevOps, logs are important to the overall function of continuous feedback. Logs need to be mined constantly to identify when exceptions are thrown so that engineers can provide a timely fix. There are many logging tools that make this process easy, using heuristics and pattern matching to identify potential security defects from exceptions recorded in log files. Logging exceptions also feeds into the alerting and monitoring tools which I discuss in Chapter 6. These can identify where exceptions are thrown due to potentially nefarious activities, such as blind injection journeys carried out by hackers.

Microservices

The concept behind microservices is to create an architecture that allows independent services to be designed, developed, deployed and maintained according to a business domain model. Each service has the capability to communicate with other services without having direct dependencies on them, and each service independently manages its own data storage, data retrieval, business logic and presentation. This architecture is in stark contrast to a traditional monolith in which all business logic is packaged into a single deployable application. These types of applications can be huge and complex. Although monolithic systems may be modularised around business domains, generally the various components are tightly coupled together and connect to one or more databases with large schemas that store data for all relevant business domains. When engineers need to make a change to a feature in a monolith, the whole application and database need to be re-compiled, tested and deployed together.

Deployments are invariably problematic. They normally consist of complex manual steps, requiring the entire application to be taken offline leading to long periods of disruption for customers. It's not unusual for operation engineers to spend long periods over weekends and evenings fixing issues during the deployment process, while development engineers remain on standby to apply quick fixes to the code. Mistakes are frequently made during these high-pressure scenarios as engineers battle with instability issues, sporadic failures or even corrupt data. From a security perspective, in their haste to push the monolith into a production ready state during the release window, engineers exercise poor practices. For example, if a service fails to start due to incorrect access policies to read data from a database, the engineer will ignore the principle of least privilege and assign the service higher permissions than necessary to work around the problem. It is common that code changes are made during deployment, bypassing normal activities such as code reviews and testing. These changes often create security weaknesses in the application.

Microservices, on the other hand, are designed, developed, tested and deployed as isolated components by a dedicated cross-functional team working as a unit. As a result, they are an enabler for the five ideals of DevOps I discussed in Chapter 1: independent and focused teams with greater productivity, psychological safety and excellent customer experiences. Teams can also achieve greater security within their value streams through low-risk releases driven by security focused design, test automation and continuous learning.

Microservice security

Microservice architecture is not implicitly more secure than monolith application architecture, although there are some benefits in a distributed architecture. Consideration must be given to the way microservices are designed and implemented to avoid introducing potential weaknesses into the application. Therefore, your engineers need to factor security into the design and development of microservices.

Security within monoliths and microservices is handled differently, so it is not ideal to copy security features used in monoliths to a microservice architecture. In monoliths, when a user authenticates by passing valid credentials to an application, a session token is generated and stored as an in-memory session ID. When the user makes calls to the monolith, the session token is validated, and the user is authorised to access the service if permitted to do so. However, in-memory sessions are not a viable option in microservices since each service runs in its own process within a distributed architecture. To overcome this hurdle, it is important to design microservices to manage authentication and authorisation independently of the services, while still maintaining the levels of confidentiality required by the application.

The best option for authenticating a user in a microservice is to centralise authentication requests within an *API gateway* rather than allowing the services to handle authentication by themselves. This reduces the likelihood of introducing security vulnerabilities into the application, and DevOps and infosec engineers can easily manage it working in collaboration. Once authenticated, an access token is provided to the user which can then be used to make a request to a service. However, another design decision that should be considered is how to authorise a user requesting a service. Using the API gateway to manage authorisation couples the API gateway to a service, while allowing services to manage their own authorisation provides a loosely coupled solution but requires the authentication token to be passed to the service, which increases the risk of it being intercepted. A neat solution for this problem is to use OAuth 2.0 within the microservice. OAuth 2.0 is a framework for managing client authentication and the access tokens and refresh tokens used to authorise the user. The access token, normally a JSON Web Token (JWT, pronounced 'jot'), is transparent – it contains the user identity and the roles of the user. To prevent the token from being used by a malicious attacker who may have intercepted it, each access token is set with a short expiry date.

As well as implementing a strong authorisation and authentication model, there are other security measures that should be incorporated into a microservice. For example, since each service communicates over the network, it is important to secure the communication channel for all APIs using Transport Layer Security (TLS) to encrypt data in transit. If services are deployed in containers, the provision of certificates can be managed via a service mesh (which I discuss later in the 'Service mesh' subsection). Another design consideration for microservices is to expose a specific set of APIs available to the public. These public APIs should be designed to incorporate rate limiting to prevent DDoS attacks and should include authentication mechanisms as described above in the 'Authentication' subsection.

Microservice architecture fits well with DevOps. It provides an easy way to develop, test and deploy features without impacting the running of applications, or it can be used within experimentation to turn on and off features for all users or specific groups of users. Additionally, automating the deployment of microservices offers opportunities to implement test automation to improve the security of the application and they allow for more granular access control.

Container technologies

The growth in *container* technologies has been driven to a large extent by the adoption of cloud hosting services; the emergence of DevOps; and the need to develop, test and deploy new products and features to customers with ever greater efficiency and lower cost. These technologies are designed to simplify the value stream and minimise disruption to customers, yet this simplicity belies the underlying complexity in the way these technologies integrate. With added complexity comes greater risk in terms of cyber security. Misconfigurations, poor design and improper implementation can leave gaping holes in the defences for attackers to exploit; therefore, it is essential that DevOps engineers and infosec understand the security implications involved in setting up a container-based service. I will describe these technologies next and provide some guidance on how to design, implement and maintain the security of products hosted in a container-based architecture.

What are containers?

In software development, containers are lightweight, small, deployable instances of applications that contain the minimum resources required by an application or service to run on a host. Abstraction of software from the underlying infrastructure allows teams to easily develop, test and deploy code to a production environment, making them ideal for applications and services hosted on cloud platforms. Containers are running instances of images which are read-only, standalone, executable packages of software that include everything needed to run an application: source code, runtime, system tools and libraries, and runtime settings. Containers can be created, started, stopped, moved or deleted. Each container is defined by its image in addition to any configuration settings applied when the container is created or started.

Containers are designed to be ephemeral, lasting for as long as they are needed. This approach is unlike traditional architecture, where applications and services are deployed to a server and nurtured by developers and operations over a long period of time. This distinction is often compared to the difference between *cattle* and *pets*. Pets are cared for individually by their owners, whereas cattle are managed collectively in herds. Given this distinction between container architecture and standard architecture, there should be no need to access containers at runtime in order to make changes to them. Operations engineers (and development engineers) are used to accessing a live service to make a quick edit to resolve a production issue; however, containers should be designed to be immutable. Immutability reduces the attack surface considerably, preventing anyone – including attack actors – from making changes to containers. This is a crucial design concept and means that remote access features to containers, such as Secure Shell (SSH), should never be enabled.

Container security

Liz Rice, in her excellent book *Container Security*, provides a checklist of the various security requirements for designing and implementing containers. This list is an excellent starting point for anyone looking to deploy containers in their organisation. There are several threats directly affecting containers which your engineers must be able to defend against, including escaping containers, exposing container secrets, and compromising container images and container hosts. The applications running within containers, as well as the connectivity between the network in which the containers are running and the organisation's network, must also be designed securely. Therefore, container security requires a number of architectural design decisions that your engineers must consider to address these threats.

Each container must be isolated into its own workspace via a set of namespaces. Isolation limits their ability to access other workloads running in other containers, as well as restricting access to the host file system and control groups. Since containers run as root, explicitly isolating them is critical to prevent malicious actors exploiting weaknesses, causing them to escape the container to access other resources.

In most cases, applications running inside containers will need to access credentials at runtime to access other resources such as databases. How they manage these secrets is a critical step in the design process.

There are various options for injecting secrets into running containers, including embedding encrypted values in the container image so that they are injected during the build process, storing them in environment variables or using the file system accessible through a mounted volume. Each of these methods has weaknesses and strengths, which means that choosing the appropriate one should be based on risk appetite and context.

Images are built using a set of command-line instructions listed in a simple file such as a *Dockerfile*. The contents of this file must be maintained and reviewed by individuals with an understanding of the security risks associated with image files. The file must also be treated as a highly restricted resource stored and protected through access controls and monitored for changes. Failure to protect the images risks malicious content such as malware being injected at build time.

Finally, containers are hosted on virtual machines (VMs), either in the cloud or on bare metal, so it is important to design these systems to minimise their exposure (such as ensuring only the bare minimum services are installed to run containers) and scan them regularly for emerging known vulnerabilities.

Containers can be a weak link if they are not designed with a strong and robust architecture. I strongly recommend that your engineers familiarise themselves with the types of threats associated with containers and know how to apply security measures to protect your organisation.

Image security

Image files describe how containers are built and run. In other words, they define any application running in a container. This means they must be treated as highly sensitive assets to prevent malicious users taking advantage of them. They must be stored securely, which I talk about in the 'Image registries' subsection. They are also prone to misconfigurations that can create security risks, which I discuss here.

Images are based on layers, the lowest of which is the base layer. This layer normally contains the operating system (OS) on which the applications within the container will run. It would be a mistake to include a full OS in the image file because a containerised application only needs the bare minimum OS features required to execute. There are a number of lightweight base images your teams should use which are designed to be small, secure and efficient, compared to fully-fledged OSs. By not using a full OS, you significantly reduce the attack surface of the running container that hackers can exploit, minimising your risk. You can also use multi-staged builds to reduce the image size, further removing the features of the base image required for the build and compile phase to a smaller subset required for running the application.

Your DevOps engineers must also consider the identity of the container at runtime when creating the image file. Unfortunately, many image files use a default user that has root access to the container host.

This means it is essential that engineers specify a user with least privileges to run the container. Because the default value is the highly permissive root user, if the engineers do not explicitly set this user, the risk of an attack on the host, which we call *escaping the container*, is significantly higher.

Another common mistake made by DevOps engineers is to hard-code secrets within the image files, either encrypted or in clear text. Since these files are human readable, it is extremely bad practice to store any secrets within them. Even if the files are not accessible to engineers without the right level of privilege, these secrets can still be revealed and exploited in a running container, described in the 'Container security' subsection. Therefore, the simple rule for your DevOps teams is: *do not hard-code secrets in the image files*.

Ensuring that images can be trusted by other teams consuming them needs to be factored into the design of the image file. Firstly, your DevOps engineers should sign their own images and put in place mechanisms to verify their authenticity. This action prevents an attacker from altering the image files between when they're deployed and consumed – a man-in-the-middle (MITM) attack. It is also advisable to design a naming convention for the image tags that help engineers distinguish between the different versions of images. This is important in identifying potentially vulnerable images from their patched versions. Engineers should name their tags with a version number, such as *alpine-v2.01*, instead of relying on the tag *latest* (numbered versions can be easily identified whereas *latest* is potentially misleading). Although this may carry overheads in automation that looks for the *latest* tag, you can overcome them with design principles that support a dynamic naming convention.

Image registries

An *image registry* is a service responsible for storing and distributing container images. Most organisations use a third-party registry such as Docker Hub, Google's Container Registry or Amazon's Elastic Container Registry. They can be public or private, their use determined by your organisation's policy on image registries: private registries need to be maintained within your organisation. Each image is stored along with its image *manifest*, which defines the layers that make up an image, and its image *digest*, which is a unique identifier of the image based on a hash of the image and its manifest. Images can also be labelled for easier identification using *tags*, such as using a version label. However, since these tags can be applied to multiple images, a tag is not a guarantee of pulling the same image from the repository each time. From a security perspective, there is an important distinction between a tag and a digest, particularly for asserting the integrity of the image.

There are obvious security measures that your DevOps teams should implement to mitigate risks associated with image registries. One of the main concerns is using an image that a hacker has compromised either at rest in the registry or when pulled from the registry. An effective control for this is to restrict engineers to access only approved registries, whether private or public. Engineers should have limited access to images within the registries based on their role within the value stream. For example, engineers should only use approved base images that have been scanned for known vulnerabilities. To guarantee the use of approved images, they should be signed with a cryptographic algorithm that can be verified at runtime.

The *Dockerfile* is an important asset for images and should be treated like any other valuable source code file – it should be version controlled and protected through access controls. There are also some security considerations to factor into the design of the Dockerfile, such as referencing trusted base images and using multi-staged builds to remove components that are not needed at runtime (such as compilers).

Orchestration

Orchestration is an automated process for managing the workloads of containers across multiple clusters. Each cluster consists of a *control plane* which controls a set of worker machines called *nodes*. The control plane facilitates activities such as scheduling, provisioning and scaling instances of containers. When the containerised service requires more resources on the host infrastructure, such as CPU and RAM, orchestration spins up more containers across multiple hosts to meet this demand; when demand drops, it terminates containers to free up the host resources, reducing costs. There are many open source and vendor orchestration tools available, such as Docker Swarm (a native clustering engine for Docker) and Kubernetes (K8s), which has become the main player in container orchestration. *HashiCorp Nomad* and *Apache Mesos(phere)* also have a small share of the market (at time of writing), and there are cloud-native versions such as *Amazon Web Services* (AWS), *Elastic Container Services* (ECS), *Google Cloud Kubernetes Engine* and *Azure Container Instances*.

Designing a secure orchestration architecture is important. A rogue user taking control of your clusters could cause a lot of damage to your organisation, so you need to build in protection from the start. Unfortunately, some of the tools – particularly older versions – are not secure by default, so familiarise your engineers with the inherent weaknesses of the orchestration tools they are using and mitigate against them accordingly. Orchestration is managed either through an API or through a dashboard. In both cases, authenticated users must be authorised to perform the necessary tasks on the cluster. Without this control, anonymous users would have full access to the cluster, giving them root access to your organisation's crown jewels. Orchestration is powerful and needs careful design and implementation to protect against malicious activity. There are tools available, including the Center for Internet Security (CIS) Benchmark checklist, that provide guidance on how to build a secure container architecture.

Service mesh

A *service mesh* is a dedicated and configurable infrastructure layer designed to extrapolate common network tasks from the application layer, making service-to-service communication safe, fast and reliable. The abstraction of network communication from the application layer allows the scalability of microservices within a cloud-native environment since it allows developer engineers to concentrate on application features and operations engineers to concentrate on inter-service communication. A standard service mesh implementation uses a proxy server to allow one service to find another to communicate with it. This specialised proxy server is often called a *sidecar*; it runs alongside the container whose traffic flow it manages. Other architectures include a *library*, which runs inside the code and is, therefore, bound to the language the application is written in, and *node agents*, which act as daemons for containers running in a specific node. The most common service mesh technologies are based on sidecar proxies.

A container architecture is ideal for a DevOps environment as it provides a high level of control over service delivery and availability. Implementing a service mesh makes it easier to route traffic during experimental feature testing with a subset of customers and allows for zero downtime when a new feature is rolled out to the entire customer base. Service meshes are also a security control in their own right since they act as a gateway for verified communications between containers.

There are several security measures to consider when working with service mesh technologies. Since the configuration of a service mesh is managed using YAML, it is important to validate the YAML file for good security hygiene. In particular, ensure traffic policies are enforced, RBAC is in place, connectivity between services is protected (using mTLS), user roles are clearly defined and service roles are bound to the correct processes. Security policies can be applied at a granular level within the service, the namespace and the mesh to explicitly enforce access controls. The service mesh can manage certificates and keys for all namespaces and enforces different access control rules to the services.

Securing the pipeline

Being able to deliver software to your customers as quickly as possible is a key driver for the adoption of DevOps practices. Developers make small, incremental changes to their products, which are processed within an automated workflow that builds packages, tests the changes and deploys the product to the relevant environment. This workflow is known as the continuous integration/continuous delivery (CI/CD) pipeline. The process starts with the developer checking in the code to a source code management (SCM) system. This triggers a build which involves extracting updated source code from the SCM, compiling the application, running a suite of automated tests and moving the package to a target environment, such as a testing server, staging server or even a production server.

With so much data moving across multiple permission boundaries, there are plenty of risks that your DevOps engineers must address to maintain a secure development environment. Having a secure-by-design architecture plays an important role in maintaining a secure pipeline as well as a secure product. Scanning small chunks of code, or scanning microservices developed with a specific business functionality, is more manageable than scanning a large monolithic application every time a developer commits a code change. Keeping application passwords secure during the build and deployment process is also a major concern for security engineers. Unfortunately, passwords stored in clear text within application configuration files – or, worse still, within the application source code – is a common problem. There are various solutions to keep keys secure within the pipeline, including key stores and HSMs. Underlying this is a product architecture that makes key management an easier proposition for development engineers.

It is critical to appreciate the value of the engineer's artefacts when developing software. Your source code is among the most important assets your organisation owns. It contains the algorithms that give your products a competitive advantage: your organisation's intellectual property. Therefore, like any other valuable asset, you must keep source code, application configuration files and deployment scripts secure at all times. Only engineers working on an artefact need access to its files. If your teams are structured around product value streams, granular access policies will permit engineers to edit and commit code for their products only. They will also restrict source code editing to engineers, while testers and product owners have permission to read the files only. Least privilege policies must allow your team members to complete their tasks efficiently while restricting their ability to make changes that are outside the requirements of their role.

Once source code has been committed to a source code repository, triggering the CI/CD pipeline, there should be no further human intervention in the journey of the source code to the target environment. All tasks performed on the CI/CD pipeline are fully automated in a DevOps environment. Applying security to this also means that only the processes that are designed to perform the task can action the tasks carried out in the CI/CD pipeline. Operations engineers configure each step of the pipeline securely by establishing trust between the various components, meaning specific steps of the pipeline can only be activated by authenticated processes authorised to perform that task. Without secure configuration, a malicious actor can use a process to perform a task that should not be permitted by that process, such as using the build task to add a malware component to the package by updating the relevant scripts to fetch the component from a third-party site. If the build task has permission to access the internet, the malicious action is easier to perform; therefore, access policies need to extend beyond human users to processes baked into the pipeline.

Continuous deployment is an extension of the CI/CD pipeline that involves pushing an engineer's code changes to production. In DevOps this can occur many times a day. The security of the CI/CD pipeline ensures that what is being deployed to production is the intended changes made by the engineers. Automating the steps and providing strict access policies for the processes involved in code deployment provides a high level of security, yet it makes separation of duties more challenging. The person writing the code should not be the person deploying it to production. In continuous deployment, deploying code is an automated process, triggered by the developer submitting changes to the source code repository. I am often asked how this impacts regulations that require these steps to be carried out by different people. The short answer is that in DevOps the developer is responsible for writing code, and operations is responsible for writing the automated process that pushes this code through the pipeline all the way to production. As long as this practice is followed and evidenced, it will fulfil the needs of compliance. The additional involvement of security engineers who build the automated security tests as part of the CI/CD pipeline lends further weight to this end-to-end process supporting segregation of duties.

Ultimately, a secure design extends beyond the architecture of the applications and services being developed for your customers. Supporting a robust operations architecture that in turn supports the delivery of software is just as important for maintaining a secure offering to your customers. Poor architectural decisions in setting up the CI/CD pipeline will lead to poorly implemented procedures that could leave your products open to security incidents. Reducing this risk requires treating the CI/CD pipeline as a product in its own right, applying the same high level of diligence as you would to the product you are delivering to your customers.

Conclusion

By following good design principles in developing application software and infrastructure software, DevOps teams can implicitly and substantially reduce the risk of introducing vulnerabilities into the products they develop and support. Building on security education strategies, engineers should design products using principles that have been established within the software development industry over many years.

In this chapter, I explained the importance of a secure-by-design strategy and then moved on to the practice of threat modelling, which allows engineers to assess risks when designing a feature within the value stream. I then explained the common weaknesses associated with poor application design and the importance of being able to identify these weaknesses within your organisation's software. I then went on to describe a number of critical design patterns that can help development engineers and operations engineers produce high-quality applications and infrastructure. In the next chapter, I introduce the third layer, which builds on the first two and allows teams to test that products have been developed securely.

SIX

Layer 3: Security Automation

Having put in place the first two layers, I move on to the third and final layer of the security foundation for DevOps. In this chapter, I introduce the concept of security automated testing. I start by explaining the importance of this layer before describing the security test pyramid, which is based on an idea that has already gained traction in Agile software development. I present various application security testing tools, explaining the different concepts behind them, and I explain why you should not implicitly trust all the code your engineers use and how you can validate the use of third-party components. I touch upon a number of ways to integrate test automation of infrastructure code before explaining the purpose of non-automated security testing and why it is important to include it within DevOps practices. Finally, I explore alerting and monitoring and managing vulnerabilities within a DevOps framework.

The importance of security automation

Testing applications and infrastructure is a critical component of any software development lifecycle. Testing takes on many forms, such as:

- Testing a new feature to validate that it does what it is meant to do
- Testing the performance of an application to assess whether new features overconsume resources such as memory, CPU or network
- User acceptance testing (UAT) that determines whether customers are able to use the new feature efficiently and accurately
- Integration testing, which ensures that different components of an application can communicate as expected, such as an API being able to talk to a database

There are also different types of security testing, ranging from simple assertions that components of the application have been configured correctly to complex penetration tests which are designed to check whether any vulnerabilities have been introduced across the entire platform. With so many tests required, it is important to fully automate as many of them as possible. It is also essential to provide feedback from testing early in the delivery lifecycle when it is more cost efficient to identify and fix issues. Feedback should also be accurate and easy to consume so engineers know what issues to fix and how to fix them as well as eliminate false positives that can waste valuable engineering time to validate.

Some vulnerabilities within an application can be discovered implicitly through various forms of functional and non-functional testing. For example, validating whether a password format meets a specific criterion is normally a test case within a functional test suite, and performance testing can identify potential performance issues that may lead to a denial of service attack if exploited. But these tests are not specifically looking for security defects and are not comprehensive enough to give infosec confidence that an application or service has been fully tested for security vulnerabilities. Security testing is a specialist area with its own set of tools and practices designed to expose these vulnerabilities.

There is a trade-off between the accuracy of manual security testing and the speed and coverage of automated testing. Cost and efficiency should govern how you incorporate manual testing and automated testing. Manual security testing is better suited to finding outlier defects, while automated testing provides greater coverage to identify common weaknesses. Traditionally, projects rely on a round of manual security testing when the project is close to going live. The cost of finding defects this late in the software delivery lifecycle is high as it often requires delays to the product delivery to remediate any high-risk vulnerabilities – or, worse, the project goes live with known defects that could not be fixed in time. On the other hand, DevOps product development does not have the concept of a final project delivery; it is based on a continuous evolution of a product. In this scenario, DevOps builds on the Agile practices in which automated testing is carried out on a continuous basis during the delivery lifecycle. As we shall see, manual tests are still relevant in the DevOps way of working, but more emphasis is given to automated security testing.

The security test pyramid

In *Succeeding with Agile*, Mike Cohn introduces the concept of the test pyramid. The pyramid represents three levels of testing: *unit tests* form the base of the pyramid; *UI-driven tests* are at the top of the pyramid; and, sandwiched between, are *service tests*. The higher up the pyramid you go, the fewer tests you have, but the cost of each is greater. A comprehensive test strategy includes tests from all layers of the pyramid, starting with a higher number of granular tests at the bottom of the pyramid and ending with a small number of broader tests at the top. It is important to avoid the ice-cream-cone antipattern in which there are many manual tests at the top of the cone and a small number of the more cost-effective unit tests at the base.

We can overlay the various security testing practices onto Cohn's test pyramid to form a *security test pyramid*. Although manual security testing has been a core component of traditional software delivery for a number of years, as organisations adopt new ways of working integrating automated testing has become a necessity. In DevOps, the use of manual penetration tests creates bottlenecks in the delivery cycle; nevertheless, security teams persist in prioritising manual testing over automated testing, failing to keep up with DevOps. This risks the security testing profile looking more like an ice-cream cone. Unfortunately, in this scenario of opposing priorities, I have seen many instances of delivery teams bypassing manual security testing completely to avoid impacting delivery schedules.

Adopting the security test pyramid is a viable solution that satisfies the demands of security teams as well as those of the delivery teams. The figure below shows the security test pyramid. As we progress through Chapter 6, I provide more detailed analysis of the testing methodologies you can include in your overall security testing strategy.

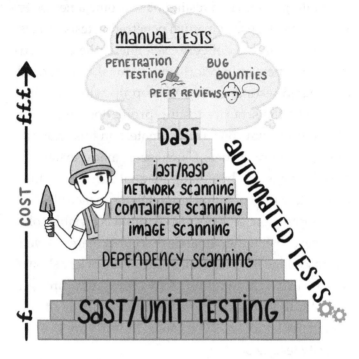

Security test pyramid: scaling test automation

- At the lowest level of the security test pyramid are unit tests and static code analysis tests. These allow engineers to test their code as they write it and fix problems within their IDE or when they commit code to the source code management tool.
- The middle level consists of several tools that work across components such as software composition analysis (SCA), container image scanning, network scanning and instrumentation tools. These tools require integration between an

application's services and components, such as databases and APIs.

• At the top of the pyramid are dynamic end-to-end test automation tools that perform exploratory testing on complete customer journeys.

Sitting above the pyramid we find a 'cloud' of testing practices that are not integrated into the CI/CD pipeline, such as manual penetration testing and bug bounty programmes. We also find more invasive manual processes such as code reviews.

The objective is not to flatten the pyramid but to maintain its structure by integrating testing practices across the delivery value stream. Unit testing and static code analysis cannot identify potential weaknesses at the integration points of the various components, so service tests and UI-driven tests are essential for detecting vulnerabilities across boundaries. Running a full penetration test involving many different components to locate easy-to-detect vulnerabilities is wasteful in terms of time and cost: you would expect to identify these vulnerabilities using tools further down the pyramid. An efficient and effective testing strategy involves minimising costs while maximising the ability to discover weaknesses in your products.

One of the core values of continuous delivery is the incorporation of fast feedback loops; therefore, automate tests as much as possible and make their results immediately available to the engineers who need to act on the test outputs. If engineers are unable to run tests and see their results for several hours after making the code changes, failed tests will take a long time to resolve, slowing down the delivery pipeline.

Test duplication should also be avoided; if you are testing for a condition lower in the pyramid, there is no need to replicate the test further up. If a test fails at the higher level but there is no test at the lower level, write the lower-level test (such as a unit test); once it has been validated, you can safely remove the higher-level test from the test suite. It is preferable to push the security tests as far down the test pyramid as you can where the costs are lower.

Application security testing

An application is the gateway between the customer and the core business function. The customer uses an application to access services or purchase products that the business offers, so it provides the greatest security risk for the business and its customers. Hackers exploit vulnerabilities in the application to target an organisation's commercial assets or its valued customers. To eliminate these risks, a company could interact with its customers through a traditional bricks-and-mortar exchange, in which the customer visits a physical site to purchase goods or access resources. This type of exchange is not a viable option in the digital age. People want instant access to shops, banks and services via the internet; the high street of today reflects this changing attitude as retail outlets lay empty and banks become trendy cafes and wine bars. In the world of traditional bricks-and-mortar commerce, security means that when the stores are closed assets are kept in safes overnight, shop fronts are shuttered and locked, and intruder alarms are activated. In the world of online commerce, security is more complex and we have seen how critical it is to develop applications with security from the outset. But how do you know that your security controls are doing what they are meant to do? Once you have designed secure systems, you need to test them, and, in DevOps, this means automated security testing. There are several options available to your teams to integrate automated security testing into your value streams.

Static application security testing

The concept of *static application security testing* (SAST) is simple: it involves inspecting an application's source code to identify any code that a hacker could exploit. Generally, there are two types of SAST tools: those that scan the raw source code before it's compiled, and those that scan the source code of decompiled executable libraries such as DLL and JAR files.

Static analysis scans inspect the source code line by line for patterns that match a possible vulnerability signature. As a result, the SAST tools tend to have excellent code coverage. Their main drawback is that each line of code is assessed out of context. The scanners ignore controls outside the scope of the scan, such as code in other libraries and external processes. Examination of the code does not always consider whether data is already in a secure state once it has reached the application, and tests would not know whether the data is validated and cleaned upstream or even downstream within the application domain. Therefore, there is a tendency for SAST to record a high number of false positive results. Most SAST tools, however, provide engineers the means to configure the scanner to reduce the reported number of false positives. For example, if logging is handled by an external library which has built-in mechanisms to mitigate against log forgery, an engineer can switch this scan feature off or reduce the criticality of the rating based on the efficacy of the existing control. Engineers must use this feature carefully as there may be a temptation to remove genuine defects from the scope of the scans purely to reduce noise or to save time when under pressure. To counter this, many SAST tools provide multiple levels of authorisation, limiting the ability to change scan configuration to a select few, such as dedicated security engineers.

When configured correctly, SAST is a low-cost defence against some common security defects hiding within source code, such as those documented in the OWASP Top Ten lists described in Chapter 5. The tools often integrate into the integrated development environment (IDE), enabling software developers to scan their code during development. Any vulnerabilities reported at this stage of development can be easily and cheaply remediated before the code is pushed into the deployment pipeline via a source code repository.

Knowing how to prioritise and fix issues that a SAST tool identifies is crucial. Since the scans target source code, when an issue is identified the tool reports its location, making it easy for the programmer to locate and fix. The tool rates each defect, normally based on ease of discovery, exploitability and its impact. Although many commercial and open source SAST tools provide elementary tips on fixing security defects within their respective dashboards, it is still vital to have an effective security education strategy in place within your organisation.

Because SAST scanners inspect source code, they need to understand the development language used within the software; therefore, it is important to choose tools that can interpret the code, whether it is Java, C#, GoLang or another application software language. This means that a single SAST tool will not necessarily cover all the development languages used within your organisation. Each DevOps team should be encouraged to experiment with the various offerings and choose those that provide the best outcomes in reducing security defects within the applications and services they develop. Their selection should be based on criteria including how effective the tool is in identifying issues, the number of false positives it raises, the ease of use and how well it helps engineers to fix security defects.

In addition to using SAST within the IDE, it should

be integrated into the CI/CD pipeline. This allows a scan to be run on code that has been checked into the relevant branch of your organisation's source code repository. This extra layer of defence provides another feedback loop to detect and fix security defects before they are deployed. The scan outputs should easily integrate with the dashboards your DevOps teams use to monitor the state of the deployment pipeline.

Stuart Gunter, a cybersecurity specialist with Equal Experts who has worked on application security testing integration for a number of clients, states that using SAST effectively is not easy, primarily because the tools do not necessarily meet the needs of high-cadence delivery teams. The high number of false positives and performance issues when integrating into a large CI/CD pipeline can be a hindrance to effective uptake of SAST tooling. Nevertheless, it can be effective when other approaches are not viable. I worked on a project for a payment provider that used SAST to meet delivery time constraints set by regulators and compliance's demands to carry out comprehensive security testing during development and releases.

Dynamic application security testing

Unlike SAST, *dynamic application security testing* (DAST) does not have access to the source code. It is a form of black box testing that attempts to detect vulnerabilities by performing certain types of attacks against a running application, usually a web application. The tests are designed to perform common fuzz testing (also known as *fuzzing*) techniques or brute-force attacks. Fuzzing is the process of injecting various types of malformed strings into a website in an attempt to make the application perform an unexpected and potentially harmful action. Although DAST is a common manual testing technique, several vendors and open source projects have attempted to automate the process. These automated DAST tools are integrated into the deployment pipeline to test running applications in a pre-production environment. By recording functional test scripts, using tools like Selenium to automate the behaviour of a user, the DAST tool can replay the resulting scripts multiple times using different predefined malformed payloads, such as SQL Injection and Cross Site Scripting (XSS). There are some potential drawbacks, though, depending on how test environments are set up. For example, well-designed testing environments mimic production enforcing the use of a random, non-repeating value called a *nonce* to prevent replay attacks; this would hinder the DAST tool's ability to run multiple payloads using the same nonce. Your DAST tool needs to be able to work with this particular application feature rather than requiring DevOps engineers to turn this feature off in the target application to allow testing to progress.

DAST's effectiveness is dependent on the amount of interaction between the web client and other services; more interactions normally results in lower code coverage. For example, websites feeding dynamic content to the user based on specific data inputs may limit the effectiveness of DAST; to counter this, your test engineers could write scripts that use different values to cover more use cases, although the permutations could be too high to provide comprehensive coverage.

Because DAST replicates real attacks – which, if successful, prove the exploitability of a vulnerability within the application – the number of false positives is much lower than with SAST. This means engineers can focus on fixing genuine vulnerabilities rather than spending time on validating them. Configuring DAST requires advanced security knowledge to set the tests up correctly based on the potential attack surface of the target application. It is worth considering inviting experienced penetration testers into the DevOps team to help configure these tools, imparting their knowledge to the team members in the process.

Many DAST tools work across multiple languages and frameworks, providing greater coverage of the application's footprint compared to SAST tools. But this comes at a cost: it is often difficult to pinpoint exactly where to apply a fix within the source code when a vulnerability has been identified. DevOps engineers need a deep understanding of the application source code as well as advanced security knowledge to fix the defects. For DAST to be effective in a DevOps value stream, the layers of DevOps security discussed in the previous two chapters of this book must be in place. Applying DAST without laying the foundations first means that the DevOps team is dependent on central security teams to provide ongoing support in maintaining DAST and interpreting their results. DAST use illustrates my point that educating DevOps engineers improves flow by allowing them to work independently of dedicated infosec teams.

Stuart Gunter highly rates DAST as an automated testing tool. When used appropriately, its low false positive output and proximity to the actions of manual penetration testing can produce excellent results. However, he does caution that engineers must be capable of configuring the dynamic testing tool for the most accurate outcomes. They also need to integrate it into an application during runtime, which makes it more difficult to integrate into the IDE; it is therefore better suited to the CI/CD pipeline, in which runtime analysis can be carried out.

Interactive application security testing

Interactive application security testing (IAST) is a runtime instrumentation tool that passively monitors traffic flowing through an application to determine whether the underlying source code is susceptible to being exploited. Although there are few tools on the market that support this type of testing, it can be effective. The application usually runs as an agent within the runtime of the underlying framework, such as a Java virtual machine (JVM). The JVM is normally initiated as a step in the CI/CD pipeline to support testing of a running application; for example, the application may be deployed to a test server and started in order to perform functional testing or UAT. The IAST agent monitors the flow of data through the running application to determine whether there are security controls to protect the data. For example, it will detect whether a string has passed through a validation method or whether a password is encrypted at source and remains encrypted until it exits the running application.

Because IAST is an instrumentation tool, it is dependent on the activity of the application while it is being monitored. If the functional tests that are executed during instrumentation do not cover 100% of the underlying source code, security test code coverage will also be less than 100%. Despite this limitation compared to SAST, IAST is less likely to report false positives because it is observing the behaviour of the data at runtime and not making assumptions about how individual components manage data locally.

IAST tools need to understand the technologies they are monitoring, so your DevOps engineers must validate that they support the languages being used by their application development teams. Because IAST tools are able to read and understand the underlying source code, they can locate and report on the lines of code that programmers should focus on to fix vulnerabilities identified by the scans.

Unlike DAST tools, IAST tools do not require any significant pre-configuration. If they find a weakness, they report it back to the engineer via their IDE or a dashboard, often with links to documentation to help the engineer fix the defect. It is feasible to set up IAST to mimic DAST by running a set of fuzz tests while the tool is instrumenting the running application. This allows engineers to see how the application manages specific payloads; however, this is not the intention of IAST and is not recommended. Although IAST is the simplest of the tools to use, it suffers the same limitation as DAST – it requires a running application in order to carry out testing, which means it is preferable to integrate it into the CI/CD pipeline rather than into the IDE. Engineers can run IAST in the background while running exploratory testing within the IDE to provide short feedback loops, but coverage is limited to the running modules within the IDE.

Final thoughts on application security testing

There are three approaches to automated application security testing: static, dynamic and interactive. All of them play a role in a successful DevSecOps framework, and deciding which to integrate into your value streams depends on a number of factors. SAST is best for identifying low-hanging fruit – the defects that are easiest to find within the source code – but it creates an abundance of false positives due to the lack of context. Supplementing SAST with DAST or IAST (or both) is worth considering.

Make sure to avoid falling into the trap of mandating the choice of one tool to satisfy all your automated application security needs. Ultimately, the decision lays in the hands of your DevOps engineers, working closely with security engineers and security champions. Through experimentation, they can determine which combination provides the best outcomes.

Mobile security testing

Testing the security of mobile applications presents its own set of challenges. Mobile applications target multiple devices, such as mobile phones and tablets, on multiple platforms and operating systems (OSs), such as Android and iOS. This complexity requires significant effort to ensure that applications remain secure on all variations of mobile devices. *Mobile application security testing* (MAST) requires combining manual processes with automation tools covering applications running on the mobile devices as well as services mobile devices consume running on the backend. It is also important to validate the security of any third-party applications or libraries mobile applications use.

There are various security requirements that are unique to mobile applications:

- They need to store secret information, such as credentials and endpoint addresses, that allows them to make secure calls to backend services
- They run outside a secure network, which means that data needs to be explicitly secured when transmitted to and from the device via wireless access points or over mobile data service

- Data should be kept secure within the context of the running application to prevent data leaking to other applications running on the mobile devices

MAST is a combination of the previous described application security testing models (SAST, DAST and IAST) that specifically targets mobile application architecture to keep mobile applications secure. It is important to integrate a full set of testing capabilities into the mobile application development lifecycle. Tools are available to provide SAST capability for the software technologies mobile applications use, such as Java, Kotlin and iOS. Some mobile-specific testing tools can test changes dynamically, although these tools normally run on device emulators within a CI/CD pipeline and may not reflect the precise capability of a mobile device.

Your organisation should run manual tests as part of the daily routine of mobile software delivery. This requires the integration of security specialists who can perform penetration testing within a DevOps environment. In particular, it is more efficient to test applications during short sprints, focusing on changes made during these iterations, than to manually test mobile applications just before delivery.

Runtime application self-protection

Runtime application self-protection (RASP) is an instrumentation technology that identifies and blocks malicious activities at the application layer during runtime by assessing whether the input changes the behaviour of the application. The main advantage of using RASP over web application firewalls (WAFs) alone is that WAFs are based on static generic rules that predict malicious activity, while RASP uses the application behaviour to determine whether traffic is malicious. There are several reasons why WAFs may not be as effective on their own. The main reason is, unlike WAFs where the context of the data passing through is unknown, RASP is configured to detect and block attacks based on the requirements of the application. Another factor is that, to support the requirements of multiple applications, WAF configuration is complicated and prone to errors that may affect the running of an application. As a result, operations engineers relax some of the validation rules, exposing one or more applications to malicious activities. Including RASP as another layer of defence mitigates this risk.

RASP has two modes of operation: monitoring and protection. In monitoring mode, RASP only reports attacks. Issues can be documented in real time within a dashboard or alerts sent to the DevOps engineers for further investigation. In protection mode, RASP can control the application's execution and thwart an attack by terminating the application or the open session in addition to reporting on the attack.

As DevOps becomes more widespread, RASP has grown in popularity. The use of instrumentation in production provides another feedback loop for engineers, which offers another option for relatively quick identification and remediation of vulnerable code. By comparison, in the traditional waterfall model the infosec team is responsible for monitoring applications in production. When an issue is discovered, security operations centre (SOC) teams spend up to several days – during which time the application or service may be offline – trying to locate the engineering teams responsible for developing the application or service so that it can be fixed and re-deployed. RASP overcomes this weakness by enabling the DevOps team to identify potential threats, fix them and deploy a more secure version within the confines of the value stream.

RASP can also be used in pre-production testing environments where the behaviour of an application can be assessed based on various different malicious inputs. This type of testing can help identify vulnerabilities before they are pushed into production. Although dedicated DAST tools are more effective for pre-production testing, your DevOps teams should not overlook this RASP functionality.

Because RASP tools are agents, they are simple to deploy and do not require changes to your applications or services. The benefit of RASP from a DevOps perspective is that it integrates into the rapid cadence of software development by continuously monitoring live systems. Being able to identify the components within the running application in which a security defect exists allows engineers to generate tests that run during an earlier stage of the software development lifecycle, creating even greater resilience.

Software composition analysis

Adding business value is the main objective of a software engineering team. They are tasked with turning business ideas into new products or services or new features to gain a competitive advantage over their organisation's rivals. The more time they spend on delivering new capabilities, the greater their opportunity to provide a better customer experience and convert more site visits into sales, so DevOps engineers tend to avoid working on tasks that offer little or no value to the business outcomes. This aligns with the DevOps ideals of flow, focus and joy.

DevOps engineers provide greater value when they avoid activities such as writing code to overcome technical challenges or overly complex workarounds to solve a problem. These coding conundrums have often been resolved by other developers and made available as open source software (OSS). For example, there may be a requirement for a user of the system to upload a document, where the processing of the content of the document provides business value but the uploading of the document does not. Rather than building new libraries to manage the document upload, developer engineers look for one that already provides this functionality and integrate the code into the application on which they are working.

At this point, your organisation's codebase contains code that was not written by your developers. You have no idea whether it has been security tested or whether it contains malicious code or malware. Worse still, many of these open source libraries depend on other open source libraries to provide further functionality. This means that there are dependency chains of OSS within your organisation's application. According to Synopsys, open source makes up 70% of application codebases, implying that most of your organisation's applications are not written by your organisation's engineers. The challenge is knowing whether these dependencies are secure and, if not, how to make them secure. Unfortunately, many well-documented data breaches have been caused by cyber criminals exploiting known vulnerabilities in OSS.

Although application security testing can identify some vulnerabilities within the dependent code, it is unlikely to identify all issues associated with them. OSS is susceptible to different types of security defects that are not all easy to find. Some are discovered within the community or by the component's creator, perhaps by a code review or from testing, and fixed as soon as possible. Others are discovered by ethical hackers and reported back to the developers through a responsible disclosure programme, giving them time to fix the code before the vulnerability is made public. Other vulnerabilities are discovered through more nefarious means, by malicious actors who use them in zero-day exploits to attack anyone using the OSS. In most cases, once the defect has been disclosed or discovered, it is assessed, rated and documented for public consumption. Known vulnerabilities may be recorded in the MITRE Corporation's Common Vulnerability and Exposure (CVE) database, and some open source providers maintain their own vulnerability databases as well.

It is essential to check application dependencies against these CVE databases and address any risks associated with a vulnerable component. There are many software composition analysis (SCA) tools available to perform this task, each with their own strengths and weaknesses. At a basic level, they all provide details of the vulnerability and suggestions for mitigating the risk, such as an upgrade to a secure version of the component or using a different component entirely. Your DevOps engineers should continually assess which SCA products are the most effective for the languages and frameworks used within your organisation. An open source component may be considered secure today and insecure tomorrow; therefore, scanning regularly is important.

Integrating SCA scans into the early stages of software development is recommended. It is cheaper to update third-party libraries during development, when it causes the least amount of disruption, than in production environments, when remediation requires an expensive round of patching (which may cause applications to stop working or services to crash). In a mature DevOps environment, patching production software is less disruptive but still carries risk. Operations staff should provide the capability to monitor production systems for known vulnerabilities, particularly as new defects are identified regularly in OSS, allowing the DevOps delivery teams to document, prioritise and carry out remediation tasks through normal daily activities.

Unit testing

Unit testing is a low-level testing framework that tests a small unit of software to assert that its functionality is working as intended. Unit tests are normally written by developers using their normal programming language. Although software unit testing has been around for many decades, it surged in popularity following the publication of the Agile Manifesto. Unit testing became a cornerstone of the Agile movement for upholding its core value of 'working software over comprehensive documentation'. One of the signatories of the manifesto, Kent Beck, was a 'big fan of automated testing at the heart of software development' and built the first version of JUnit in 1997 specifically for the Java language. Since then, the *xUnit* framework has been extended to support many common development languages. xUnit tests allow developers to write, organise and run unit tests to validate the output of a function against expected values. Kent Beck is also credited with creating the *eXtreme Programming* (XP) methodology, which expanded the use of unit tests through the invention of *test-driven development* (TDD). TDD consists of three simple steps:

1. Write a test for the functionality you want to write. The test will fail as there is no functional code to test at this point.
2. Write functional code until the test passes.
3. Confidently refactor the code to make it well structured.

As engineers cycle through these steps, they build an application that has functionality supported by a full suite of unit tests. When developers change features, they edit the unit tests first to reflect the new functionality before writing the functional code. Because each unit test provides a use case for the feature being tested, it implicitly documents the underlying application or service. Additionally, code that has been developed with unit tests is often cleaner and better designed than other code; this is because developers, in writing the absolute minimum to pass a test, are forced into adopting secure-by-design principles.

Security engineers or security champions should also be involved in this process by helping the developers write unit tests that assess the security of the application or service. They could write unit tests to check that functions with string parameters validate the input for injection attacks, or they may produce unit tests that ensure secrets are encrypted correctly. Unit tests can also be used to validate defects that other types of testing used by the DevOps engineers have identified; this involves the developer writing a unit test that replicates the security vulnerability and refactors the functional code until the unit test passes. This increases the chance that the security defect is detected earlier in the lifecycle than during DAST and IAST validation, and it reduces the number of false positives normally associated with SAST tools.

Not only is unit testing integrated into the IDE, it also plays an important role within the CI/CD pipeline. Unit tests are run whenever a build is triggered, usually when a developer checks in source code or when the build is scheduled. When unit tests fail, causing the build to fail, developers are unable to check in new code until the build is fixed, and, ideally, the DevOps team works as a unit to fix the build.

DevOps engineers must fully commit to writing a comprehensive suite of unit tests to reap their full benefit. Teams that work under pressure often remove failing unit tests or do not write them at all in order to maintain successful builds. This is a false economy because if a failed unit test covers a security defect, the resulting vulnerability will become a bigger problem when it is promoted to a live system.

Your organisation's DevOps team may be supporting older products that have limited or no unit test coverage. Particularly if your company is transitioning from project-based delivery to product-based, there is a tendency to try to retrofit tests into the software delivery lifecycle. I believe this is not ideal. Older applications have likely been poorly designed and developed, making it difficult to incorporate unit tests into the value stream. I recommend that software engineers only write unit tests for features that they are currently working on, rather than writing unit tests systematically to increase code coverage. Although the initial number of tests will be small, over time more unit tests will be written as more of the legacy code is changed, providing greater code coverage. A side benefit of this process is the steady improvement in the quality and security of the legacy product.

Unit tests are the cheapest option and provide the greatest accuracy, but only when your organisation is committed to a strategy of TDD. You may consider writing unit tests an overhead, yet they provide the most cost-effective way of automating the testing of functional and non-functional security features. Unit tests, combined with SAST, DAST and IAST provide powerful lines of defence against simple security weaknesses.

Infrastructure as code testing

Traditionally, organisations have provisioned their own hardware, onto which they install operating systems and configure services to support their business value applications hosted within these environments. The arrival of cloud computing has changed this model. Instead of procuring and hosting hardware in their own data centres, organisations can take advantage of cloud service providers who manage the hardware on their behalf. There are three main types of cloud services:

1. *Software as a service* (SaaS), in which third-party applications are hosted in the cloud and accessed by users through a browser (eg *Google Docs*)
2. *Platform as a service* (PaaS) provides a platform consisting of an infrastructure, OS and services that can be used to build software (eg *OpenShift*)
3. *Infrastructure as a service* (IaaS) offers the same capabilities as an on-premise data centre without the need to physically maintain the hardware, meaning that organisations are responsible for the OS, middleware and data hosted in the infrastructure

There are other services, such as *function as a service* (FaaS) and *cryptography as a service* (CaaS), although they are considered subsets of PaaS or SaaS. In DevOps, where engineers are responsible for managing application software and infrastructure, the adoption of the IaaS model has become widespread. Traditional and Agile development teams, on the other hand, would benefit from the PaaS model, where the vendor manages the infrastructure, leaving developers to concentrate on delivering software.

A key component of DevOps is applying principles of software development to create and manage infrastructure using code that can be managed like application source code: it is versioned controlled, tested and deployed within an automated CI/CD pipeline. This has led to the widespread adoption of *infrastructure as code* (IaC). In traditional provisioning of infrastructure, security teams managed security, including firewall configuration, access control and identity management. This would entail operations engineers sending detailed documentation to the various security departments to review and approve the infrastructure design before allowing a change to take place, sometimes in the development and testing environments as well as the production environment. If there was a problem, the design would need to be resubmitted and go through the change process all over again. Once approved, the changes were manual, following a set of written instructions, and often had to be carried out during a change window in unsociable hours to avoid major disruption to customers. This process would take several days or weeks, which hampered the ability to deliver new products in a timely manner.

IaC has changed this model. Infrastructure code can be written, tested and deployed in just a few minutes. Although there is an obvious benefit of speed, the major problem with this approach is that it can propagate errors or security issues to many servers quickly. Security engineers should review the IaC files, but they may not have sufficient understanding of the IaC technology and the manual review would take a long time. The solution to this is to automate the testing of infrastructure code, particularly for potential security flaws.

At the time of writing, code analysis tools for infrastructure code are not as mature as tools for application software discussed in the 'Application security testing' section. However, there are a few open source and vendor products that provide this capability for a number of infrastructure coding languages such as Chef, Puppet and Terraform. Many of these tools look for common misconfigurations, such as weak or invalid encryption algorithms, incorrect security groups and overly permissive access controls. They can identify if all ports are open, if any IP address is accessible and whether access is provided to all resources. These tools can be integrated into the IDE where operations engineers write their IaC scripts, and they can be integrated into the CI/CD pipeline where build failures can block the code from being promoted to a production environment. Ultimately, this provides excellent security as well as enabling the higher cadence of the value stream.

Dynamic analysis of infrastructure code (discussed later in this chapter) scans the network to identify potential security weaknesses and implements monitoring and alerting to detect behaviour that may indicate a weakness in the infrastructure.

Container image scanning

As discussed in Chapter 5, the use of containers to package software enables DevOps engineers to develop and deploy applications within a self-contained environment consisting of the minimum features and services needed to run the application. We also saw that the availability and configuration of containers is managed via orchestration tools, while a service mesh manages the communications between containers and with the host network. Within this architecture, containers with vulnerable code or those that have been infected with malware can propagate quickly, significantly increasing the risk of a security breach. Incorrect configuration of the image file could lead to attackers breaking out of a compromised container to access other containers or even the host OS running the containers. There are several tools that integrate into the CI/CD pipeline, or that can be used by engineers from their development environment, to validate the security of your containers.

Image scanning tools allow engineers to verify the security of an image based on information provided within the image file. It is important to configure the images securely, making sure that they use a least privileged user, do not contain hardcoded secrets, use safe images and are securely stored. Scanning images involves validating configuration settings to ensure the attack surface is minimised (for example, checking that SSH daemons are not included in the image), identifying utilities that could pose a risk (such as a bash shell), and checking that the correct (secure) image version is not being overwritten with an insecure version.

I recommend scanning container images throughout the delivery lifecycle. Programmers should scan their own images when they are created and before they store them in the local repository. They should also scan any images they pull from external or internal registries. Operations engineers should enable regular scanning of images within the local registry and set up a deployment pipeline in which failed image scans are either flagged for review or block the build from progressing entirely.

The criteria for blocking the progress of failed images needs to be considered carefully. As an example, if an image contains a vulnerability that already exists in a production environment (such as a recently discovered CVE), blocking the build does not reduce the security risk of the production environment. Furthermore, if an image contains a critical change required for business or regulatory reasons, blocking the image due to an existing CVE could lead to further complications, such as fines or lost business. You must be able to weigh the risks and decide how to deal with failed image scans within the CI/CD pipeline.

Scanning running containers in production could identify behaviour indicating a potential attack is in progress; for example, an activated SSH session suggests an attempt to connect to a container remotely. In these cases, it is better to fix the problem within the image (such as removing the SSH daemon) and replace all affected containers as part of the continuous delivery process. It is inadvisable to try to remote into the container to fix a problem or patch the containers directly, while it is advisable to maintain immutable images. Therefore, when a vulnerability is discovered in a running container, it is preferable to create a more secure image for deployment rather than try to fix the running container. In this way, a compromised container can be quickly destroyed and replaced with a clean version before any further damage is inflicted.

Container technologies, such as the orchestration tools, are also targets for hackers. They will try to use known vulnerabilities from the CVE databases to attempt exploits against unpatched versions of these tools. Therefore, it is important to monitor the container ecosystem to detect new vulnerabilities in the versions of these tools you are using in your organisation. This should be integrated with a robust patching policy to ensure that these products are updated when new versions are available.

Image scanning is still evolving at the time of writing. Open source tools include *Clair*, *Anchor* and *Trivy*, and commercial tools include *AquaSec* and *Twistlock*. You can also find resources from the Center for Internet Security (CIS) that provide security checklists for technologies, such as the CIS Benchmarks for *Kubernetes* and *Docker*.

One of the problems associated with scanning images is that it is impossible to determine from the static file how the application behaves at runtime. An image file may include an apparently safe executable, which is activated when the container starts, embedding another malicious file (such as malware) into the container at runtime, exposing your system to attackers. A solution to this involves validating the container at runtime using dynamic threat analysis.

Dynamic threat analysis

Scanning images using automated tools within the CI/CD pipeline is an essential piece of the security jigsaw; however, there is no guarantee that the image will run as expected when it is activated. You can scan your organisation's containers at runtime in the production environment, but by the time you've detected a security issue, some of the damage may have already been done. To overcome this problem, a new set of tools that can validate the image as a runtime within the CI/CD pipeline is coming onto the market. At the time of writing, this process is called *dynamic threat analysis* (DTA), and Aqua Security has already released a DTA tool. The idea behind DTA is simple: instead of running the container in production, a sandbox is created in which to activate the container and scan the environment for unusual behaviour that indicates potentially malicious activity. It can, for example, identify unusual network behaviour that indicates possible attacks taking place, such as malware running in the container. Using DTA allows engineers to identify hidden risks that cannot easily be detected by traditional image scans before they end up in production.

Network scanning

All software is deployed into an integrated network of computers. From desktop computers to mainframe servers, and from routers to printers, every piece of hardware is connected to other hardware either directly or indirectly. This integration is what makes life so easy for us. Not only can we use the internet to do our shopping, banking and research, we can also use it to adjust the heating and lighting in our homes or answer the doorbell, even when we are miles away. Within your organisation, there are plenty of interconnected devices, from laptops and routers to printers and smart card readers. Connectivity can be physical (using cables to chain components together) or wireless (such as Wi-Fi routers or Bluetooth).

Within this interconnected super-highway, in order for messages to be routed correctly, each device uses a system address known as an *IP address*, which is a unique identifier within a network that allows these devices to communicate with each other. The most common and oldest IP in use today is IPv4, which is a 32-bit address divided into four octets of 8 bits, each represented as a number from 0 to 255 and separated by a dot (eg 198.51.100.34). Due to the huge number of devices now connected to the internet and within private networks, and the limitation of the available IPv4 address range (approximately four billion), a new IP version is available. The new IPv6 is made up of eight segments of 4 hexadecimal digits, each separated by a colon (eg c689:a275:99ef:ce0f:8f73:6438:e19f:61eb).

There are also 65,536 TCP (transmission control protocol) and UDP (user datagram protocol) ports for each IP address, allowing multiple services and applications to communicate across networks using a common network protocol. With so many IP addresses and network ports, the risk of a rogue application exfiltrating data to a specific port and IP address can be significant if left unchecked. This is where network scanning comes into play. System administrators use tools that scan networks, such as the free and open source Nmap, to determine the availability of hosts on the network, the types of services offered by the hosts, and the OSs that are running. The scanning tools can also target specific firewall rules that may be in place on the network.

Network scanning allows system administrators to maintain an inventory of their ecosystem; it also allows hackers to build up knowledge of a system to determine attack vectors they can use. This makes network mapping essential for cybersecurity to identify potential 'holes' in the perimeter defences that may allow an attacker in. Within a cloud platform – governed by shared responsibility contracts between cloud providers and consumers – the infrastructure (including the physical network) is the responsibility of the cloud provider. This means that the cloud provider is responsible for the hardware infrastructure, including all the IP addresses in its remit. The consumer is responsible for securing the network connectivity within the network instance allocated by the cloud provider, such as AWS VPCs (virtual private clouds), Google VPCs and Azure VMs; however, if your teams are deploying to an on-premise infrastructure the security of the whole network is your organisation's responsibility.

Network scanning is not a process that sits easily within a CI/CD pipeline. Unfortunately, if your DevOps engineers scan the network every time they check in code, or when the service or product is deployed, the network is likely to start underperforming. If network scanning is to be integrated into DevOps value streams, it is more beneficial to schedule scans at regular intervals such as daily or weekly and use feedback loops to create backlog tickets to investigate and fix any anomalies. Infrastructure as code, by contrast, allows engineers to create a network topology using scripts such as Terraform and Ansible. As with application code, infrastructure code can be scanned using static analysis tools that alert engineers when potential vulnerabilities are present. These tools are not as mature as their application security scanning counterparts, but your engineers should be able to research and experiment with a number of them to integrate into the CI/CD pipeline.

Some testing cannot be automated

Automation is a key component of DevOps, but manual security testing can also be an essential part of the process. In some regulated industry sectors, it is mandatory for an accredited third party to carry out penetration tests on an organisation's online product or service. Normally, these penetration tests are carried out in a pre-production environment, although they can also be carried out on a live system. Despite the many layers of automated testing within the value stream, it is impossible for it to be 100% secure. In the worst-case scenario, a security defect that remains undetected through the delivery lifecycle and is promoted to a production environment is exploited by an attacker. This can be devastating, resulting in financial loss, reduced confidence among customers or even sanctions by regulators.

Although manual penetration tests identify the more difficult-to-find vulnerabilities, some weaknesses still go unnoticed. These can be the most damaging to your organisation and your customers, so many organisations offer bug bounties, which are incentives for ethical hackers to discover and responsibly disclose any vulnerabilities that they encounter. Although payouts for a bug bounty programme can be high, they are more cost effective than being compromised by a hacker. For example, a vulnerability could facilitate a ransomware attack that may require your organisation to pay a huge ransom and also spend a large sum on the work required to bring your systems back online.

Unfortunately, third-party penetration testing and bug bounty programmes are manual processes that are not easy to integrate into the normal day-to-day activities of DevOps practices. Nevertheless, when issues are discovered, DevOps provides the framework to fix bugs quickly and efficiently with minimal disruption to customers, and penetration testing and bug bounties can work effectively with DevOps practices.

Penetration testing

The *National Cyber Security Centre* (NCSC) defines penetration testing as 'an *authorised* test of a computer network or system designed to look for security weaknesses' (emphasis in original). During penetration testing – or 'pen testing' – authorised testers try to exploit publicly known vulnerabilities and common misconfigurations using the same techniques and tools used by an attacker. The output of these tests is a report containing a list of the vulnerabilities that have been identified during the test. Each vulnerability is given a score based on its criticality, allowing organisations to prioritise a fix for each issue. Penetration testing is often carried out as a *black box* exercise, meaning that the tester has no knowledge of the system other than the public-facing user interfaces such as websites and mobile applications. However, pen testers may request some basic information to increase their efficiency, such as a set of valid credentials to gain access to systems or a topology of the network to make navigation around the network easier. This type of testing is known as *grey box* testing as the pen tester has limited knowledge of the system. Black box and grey box testing are meant to give pen testers a similar experience to that of a real hacker by restricting what the pen tester can see. With *white box* testing, the pen tester has access to everything, including source code, valid credentials, network topology and IP addresses.

In traditional waterfall deliveries, pen testing normally takes place near the end of the project, often when fixing issues has become too expensive or risks delaying the project delivery. In the worst-case scenario, it is too late to fix issues and the product is released to live in a vulnerable state. In a DevOps environment, the various automated testing processes are designed to identify most, if not all, of the vulnerabilities within the system as early as possible. This is often referred to as *shift left*, although I prefer the phrase *expand left* because security should be integrated throughout the lifecycle by expanding from the right to the left of the delivery pipeline.

In DevOps, there are multiple releases a day; therefore, running manual tests just before the system is deployed is not feasible. Instead, manual penetration testing is carried out as part of an annual review to satisfy regulatory requirements, and as a verification process that highlights weaknesses in your automated security testing processes. Unfortunately, it is impossible to completely eradicate security defects, especially those that are complex to find and equally complex to exploit; so, when manual pen testers discover issues, you need a strong feedback process in place to manage these issues within the value stream.

The reports generated by penetration testers are static documents (they should be treated as highly restricted information), so the vulnerabilities must be documented in the normal defect tracking system, where they can be prioritised and fixed by the DevOps engineers. Automated tests should be configured to identify the same vulnerabilities so that fixes can be verified as part of the normal delivery process and reduce the risk of re-introducing the same vulnerability at a later date. Within DevOps, engineers have the confidence to fix these issues and push the clean code to production as part of the normal delivery lifecycle, often through multiple daily releases. Therefore, even when manual pen testing identifies issues, they can be fixed and pushed to live efficiently.

Bug bounty programmes

No matter how much testing is completed during the delivery lifecycle, production systems still have vulnerabilities that can be exploited. Some are known by the organisation developing the system, where risks have been accepted or controls have been put in place to mitigate the risk; however, persistent hackers determined to find a way into a system to steal data, disrupt services or commit fraud are skilled in finding and exploiting these weaknesses. Hackers may not initiate an attack directly on an application; they can exploit human weaknesses through targeted phishing attacks or social engineering to gain access to a system – either physically, such as an office or data centre, or digitally, using stolen credentials. Automated testing and manual penetration testing do not provide the flexibility and imagination used by skilled and experienced hackers to attack a system, and sometimes the first time an organisation learns about a vulnerability in its system is when the attack has already occurred. The process turns into a damage limitation exercise to restore services and compensate customers, while trying to understand how the attack happened in the first place. This may involve bringing in trained cyber-security forensic experts at huge cost to the company. In many cases, the financial loss is compounded by large fines from local regulators for breaching compliance. In short, an attack by a hacker can be devastating to an organisation as well as expensive.

A solution to this problem is to organise a bug bounty programme. Bug bounty hunters, often referred to as *ethical hackers*, are skilled and experienced cyber-security specialists who are motivated to find security defects within a service and disclose them responsibly to the affected organisation for a reward. It is important to set the level of reward correctly – if you give high rewards for easily discoverable vulnerabilities that could have been found during automated testing, bug bounty hunters will be incentivised by the easy pick-ings and the cost of the rewards will be substantially higher than the cost of finding these issues earlier in the lifecycle. On the other hand, setting a reward too low for vulnerabilities that are difficult to detect will deter bug bounty hunters from making a significant investment in time to engage with the programme, paving the way for attackers seeking greater rewards to find them.

The output of a responsible disclosure is a report that documents the steps to reproduce the attack and a description of the fix or fixes to remove the weakness. This document contains a lot of sensitive information and must be treated as highly confidential to prevent it from falling into the wrong hands. However, as with the penetration testing report, the documented issues should be factored into automated testing processes and the fixes added to the engineers' task management tool, prioritised and fixed accordingly. The DevOps processes allow changes that address these vulnerabilities to be made to the application code and infrastructure code, tested and deployed to production without any disruption.

During a bug bounty programme, defects can be identified at any time. This is in contrast to automated testing and pen testing, which take place during a set timescale. Therefore, it is important to set aside time for DevOps engineers to work on defects that bug bounty hunters report to the organisation. Prioritising tasks to address these issues within the day-to-day activities of your DevOps teams prevents the bug bounty programme from disrupting normal flow of work, particularly if the defects are critical.

If you choose to set up a bug bounty programme, you need to publicise it and provide a feedback loop from the bug bounty hunters that integrates with your DevOps processes. Failure to provide a means for bug bounty hunters to safely disclose a vulnerability to your organisation is a recipe for the programme failing. If bug bounty hunters are unable to gain traction in disclosing a vulnerability to the affected organisation, they may disclose it to the public; their motive being to raise awareness of the vulnerability and protect customers.

Monitoring and alerting

Being able to see what is happening in a live environment in real time provides valuable insights into the security and performance of the services running in production. Gathering and analysing detailed information can help DevOps engineers fix defects, increase performance and improve the customer experience. There are five components of a *monitoring and alerting* system:

1. *Data collection* – the application and infrastructure are configured with instrumentation to collect useful data.
2. *Data analysis* – data from multiple collection points is processed in real time by services to identify potential security incidents or fraudulent activity.
3. *Data storage* – data is stored centrally in order that downstream systems can retrieve it.
4. *Alerting* – data indicating a potential incident triggers alerts sent to the relevant notification endpoints.
5. *Dashboards* – data is presented to DevOps engineers and business stakeholders in clear visual displays, enabling them to identify trends and anomalies.

It is important to identify which data to collect. Too much data becomes overwhelming, whereas too little data leaves holes in your understanding of events. In both cases, the risk of missing a security incident is high. Data is collected from different sources, each providing valuable insight into various parts of the system that could prove useful if your systems are compromised.

Alerting and monitoring is an essential component of DevOps providing feedback loops that allow engineers to quickly identify and fix issues, including potential security events affecting hardware, OS, VMs, middleware and applications.

If your organisation has yet to migrate to the cloud, and your services are hosted in a dedicated bare-metal data centre managed by your organisation, it is important to monitor hardware for potential anomalies and events affecting underlying mechanical components, such as hardware failures, power-offs and cabling defects. If the services are running in the cloud, these types of alerts are not relevant as they are monitored by the cloud hosting providers. Hardware events that may indicate abnormal behaviour include an increase in fan activity, suggesting that the machine is overheating. This could be the result of the machine being used in a resource-intensive cryptocurrency mining attack.

The health of VMs (if hosted on premise) should be monitored to ensure they are running as expected. In particular, collect data relating to availability, health, CPU usage and memory utilisation that provide important information about performance and security. This data usually needs to be collected using third-party components. Your engineers should be on the lookout for any increase in CPU cycles or memory usage that may suggest the VMs are being used for resource-intensive activities, such as the aforementioned cryptocurrency mining attack or as part of a botnet. It may also indicate a DDoS attack against your system.

Whichever OS you use, monitoring data relating to memory usage, disk space and process activity provides metrics on the health of the system. Modern OSs provide interfaces that allow this data to be easily consumed, and there are many commercial and open source tools available that extract this data in real time. DevOps engineers can configure these tools to provide a response based on specific events that may indicate your systems are being attacked

Monitoring *middleware* services running in an environment provides information on thread pools, database connection pools, and other metrics that can give operations a view of the health of the middleware layer. Your DevOps engineers may identify potentially abnormal behaviour, such as changes to database schemas, non-essential active services or possible race conditions consuming threads.

Finally, *applications* should be designed and written in such a way that they provide hooks into the state of the application at runtime. These metrics are useful for business owners, who can determine how successful the application is running from a commercial perspective, as well as the DevOps engineers, who are interested in knowing the state of the application at runtime. For example, developers can determine whether the application is functioning properly by reviewing the number and type of exceptions being thrown, the health of running services, and any performance or latency issues. Applications can be configured to log events by type. It is common for modern programming languages to provide log levels such as *FATAL, ERROR, DEBUG, WARNING* and *INFO*. Your DevOps engineers should be familiar with these log levels in order to provide accurate information about the state of the running application. For example, a *WARNING* message would be used if there are failed login attempts from a single user, *ERROR* would be used if the login service fails for a single user, and *FATAL* used if the login service fails for all users possibly indicating a significant security event is happening. Engineers need to agree on the level of logging for each event type – DevOps engineers would not want to be woken at 3am to be told that a user entered the wrong password on small number of occasions, nor would they want to be informed many hours after the login service stopped working due to a persistent brute-force attack.

As you can see, there is a huge amount of data being collected. Analysis of this data is normally carried out by commercial or open source tools, frequently using heuristics to determine which data may contain anomalies such as security events or fraudulent activity. The analysis may also reveal details indicating potential commercial failings, such as the number of transactions cancelled. This is useful within the value stream in which experimentation is used to identify which features offer the best customer outcomes.

The data should be stored in a retrievable format for a period of time normally determined by regulatory requirements. There are two types of storage: short-term and long-term. Short-term storage allows engineers and business stakeholders to quickly retrieve data that give indications of how the product is performing in near real time. This information is kept in data stores for a few days and is accessible through dashboards and other services such as messaging services. Archiving data into long-term storage facilities maintains it for historical purposes and frees up resources from the short-term data stores. Some industries and jurisdictions require data to be stored for a number of years so that historical activity can be retrieved. There may be evidence within the stored data that a fraud was committed in the past but not identified at the time, and forensic analysis of this data could be used in a criminal investigation.

Dashboards and alerts are used in real time to view the current state of products and the services supporting them, and notify individuals or groups of a specific event that may need further investigation. The components function as a composite tool. When engineers receive an alert, they will know what type of alert it is but not necessarily the context. The dashboard gives engineers a view of the relevant data over a period of time before and after the alert that may give clues about why it was sent. If engineers receive an alert indicating that a threshold of failed logins has been reached, they won't know the context until reviewing the dashboards, which may indicate that the number of failed logins occurred immediately after the engineers upgraded the login service. Further studying of the dashboard reveals that a dependency used by the service is failing sporadically, and now your DevOps engineers can quickly identify the problem, fix it and deploy a working solution.

Alerts take many formats, sometimes dependent on their severity and type. A *FATAL* alert may come in the form of a text message or an automated phone call received by a small group of on-call engineers. Less urgent messages may be sent via email or appear in a Slack channel. Alerts must be used appropriately; your engineers will not be pleased if they receive multiple automated phone calls every night. In fact, they are more likely to turn their phones off to guarantee a good night's sleep! *'Alert fatigue'* can occur when an individual or team receives so many alerts they become desensitised to them, causing your engineers to miss potential security alerts. Your DevOps engineers and the central cybersecurity teams must work together to design the alerting policy that produces the best outcomes.

Dashboards are only useful if they are visible and convey information easily. They should not contain verbose stack traces or show thousands of data points in detailed data structures such as tables. Instead, data should be presented graphically with colour coding that indicates the health of the system. A traffic-light colouring system is intuitive and recognised by most people; red indicates an error causing a critical state, amber an error causing a non-critical state, and green a fully operational state. In the login failure scenario, if the number of failed logins reaches a certain threshold, a simple table showing the number of failed logins will turn from green to amber or red depending on the thresholds reached.

Integrating monitoring and alerting into a DevOps environment requires collaboration between your DevOps engineers and cybersecurity teams. Each specialty has its own set of requirements: cybersecurity specialists are interested in identifying events that may indicate that an attack is occurring, customer data is at risk or malware has been identified; DevOps engineers want information that indicates that the infrastructure and applications are running correctly. In addition, DevOps teams benefit from events that indicate how customers are reacting to a new feature or service, and whether they are more or less likely to use the feature. Traditionally, developers, operations, cybersecurity and business stakeholders worked in silos, meaning that the data reporting was inconsistent and often left the operations teams responsible for all types of alerts. In DevOps, everyone in the value stream manages alerts collaboratively to ensure the best outcome for the team and the customer.

Although alerting and monitoring are not strictly speaking test automation, they should form part of the overall test automation strategy. Being able to discover issues in real time is an attribute of DevOps that favours improved security. Through the correct configuration settings and refinement of those settings, monitoring provides an early warning of potential security issues within the production environment. Developers and operations engineers are better equipped to proactively fix issues before they become bigger problems and can react quicker when the criticality of those issues passes a specified threshold. Agile ways of working, in which changes in development can be promoted to a production environment within a short time frame and normally within business hours, provide a platform in which production issues are handled as part of the day-to-day activities. Production issues need not result in escalations, disruption to the business and emergency war rooms: they are just another feedback mechanism into the work of DevOps engineers, resulting in a more secure system and happy engineers and delighted customers.

Vulnerability management

Managing vulnerabilities is a critical process within your organisation. Automated and manual testing will identify a number of security defects and each one must be carefully recorded, rated and prioritised so that it can be resolved according to your company's cybersecurity policies. There may also be regulatory requirements that the vulnerability management system must meet. Let us explore how DevOps and cybersecurity can work together to manage security defects.

Vulnerability database

Whether vulnerabilities are discovered through manual penetration testing, automated security testing, code reviews or even by bug bounty hunters, they must be recorded in a system that can be accessed by the engineers who can fix the issues. Some traditional vulnerability management systems are managed by a central cybersecurity department who maintain the vulnerability data store under lock and key at all times. Reports of the vulnerabilities are generated when they are discovered, and the reports are sent to the product owners or project managers responsible for fixing them. These reports are likely to be PDF documents, spreadsheets or paper copies. In many cases, the reports are produced many weeks after the code was written and, in some cases, after the project team has been disbanded. If (rather than when) an issue is fixed, the engineers responsible for developing and testing the fix have no direct access to the vulnerability database to update it. Instead, they have to contact the central cybersecurity team to update the status of the vulnerability on their behalf, or, in many cases, the fix is not reported back to the central cybersecurity team at all. The updated application remains unchecked until another round of testing is scheduled. Sometimes, the vulnerability remains in production with potentially disastrous consequences. In the meantime, if another bunch of issues is introduced to the application the whole cycle is repeated.

Unfortunately, the model I have just described is probably familiar to you, especially if you have experience of working in large organisations. These practices are in conflict with DevOps ways of working which rely on fast feedback mechanisms, continuous testing, and continuous integration and deployment. The engineering team needs access to the vulnerability database managed by the central cybersecurity team so they can assess, prioritise and assign each vulnerability to an engineer. Engineers must also update the vulnerability database with evidence that the defect has been fixed, but often they have limited access to do this and rely on convoluted feedback mechanisms involving clumsy email communications to have the vulnerability database updated for them. This is a bottleneck that hinders the flow of work in a DevOps team, and DevOps engineers and cybersecurity teams become disjointed: DevOps engineers do not report fixing vulnerabilities to cybersecurity teams and even avoid notifying the security departments when they discover a vulnerability. Instead, they manage their vulnerabilities using their own defect tracking systems independently of those used by central cybersecurity teams. In the meantime, the security departments erroneously believe their database is the single source of truth of the organisation's security risks.

To overcome this problem, DevOps engineers must work closely with cybersecurity engineers to define the processes for managing security defects, whether they are discovered by automated or manual testing. This process must completely satisfy the requirements of the groups involved. This means it needs to incorporate fast feedback loops so engineers can fix the issues as quickly as possible, appropriate access rights enabling the various engineering and security team members to do their jobs while protecting data from others, and also add reporting mechanisms that provide accurate views on the security status of your organisation. Ultimately, the vulnerability management system must provide the best outcomes for your customers, who expect your products to be secure and meet the regulatory requirements of your local jurisdictions.

The security of a vulnerability management system is critical. Each defect documents the vulnerability with detail to allow engineers to locate the defect in the application or service, know the state of the application when the defect was detected (that is, the specific data used) and the sequence of events that led to the defect being found. In other words, details of security defects that have yet to be fixed are playbooks that hackers can use to locate potential weaknesses and exploit them. Therefore, access to the vulnerability database must be restricted to the relevant groups and carefully audited for unusual activity.

Vulnerability reports

Vulnerability management systems must be transparent; it must be possible to generate meaningful reports from the vulnerability data store as well as maintain a live status of current vulnerabilities. The key metrics that should be tracked are:

- *Mean time to fix* – Understanding how long issues take to be fixed and deployed to production helps to determine the current risk profile of live systems. The longer issues remain open, the greater the likelihood they will be discovered by a hacker and exploited. It is worth reporting this value by the severity level of the security defect, giving even greater insight into how quickly issues are being addressed. This is important because lower-rated issues, if not managed correctly, can remain open for much longer – in some cases, indefinitely. This information may also highlight potential skills gaps among the engineers if a specific type of defect takes longer to fix than others. The mean time to fix should also be tracked over time for any significant changes.
- *Most common defect categories* – By categorising security defects, you will see which defect types are more common within your organisation's systems. Recording this over time can indicate whether there are any emerging risks that need to be addressed. As I discussed in layer 1,

educating engineers on how to develop secure systems is important. However, entropy can set in, meaning that knowledge on mitigating some security risks is lost over time. Identifying when the frequency of a particular issue type starts to increase can help determine the cause and allow teams to adjust accordingly. Likewise, investment in education can be directed towards mitigating security defect types that are more common. Some categories can change after functionality has been added or removed, or following the adoption of new technology or changes in components used during the development lifecycle. The report may show that certain categories of issues remain prevalent over long periods of time despite the engineers having the knowledge and skills to fix them. This may indicate a prioritisation issue, in which features take precedence over certain categories of security defects. In this case, I suggest you focus on reducing the number of open issues of a particular type as a priority.

- *Severity scores* – Knowing the severity scores gives you a clear indication of the risk your organisation is exposed to. Measuring this over time helps identify trends in the overall risk so that you can act accordingly. For example, initial reports may indicate a high number of medium-rated issues and a handful of low-rated issues. Over time, this may change to a higher frequency of low-rated issues and a small number of medium-rated issues. This could indicate that, although medium-rated issues have been prioritised over lower-rated issues, more medium-rated issues are fixed at the expense of addressing low-rated issues. Unfortunately, this is not ideal since a high number of low-rated issues accumulating over time can put your organisation at greater risk.

- *Vulnerabilities by application or service* – Most companies have a list of their assets, including their products and systems, facilities and people. This is commonly known as the configuration management database (CMDB), the purpose of which is to define and monitor the relationships between the various components of an organisation. The vulnerability management system should be linked to the CMDB so that you can track the security posture of the systems developed by the organisation. The association between vulnerabilities and related assets can help drive business decisions; for example, a product that is not core to the business and has multiple vulnerabilities could be a candidate for deprecation, while a legacy core product that has multiple vulnerabilities and may no longer be funded for development can be problematic. In the latter case, a decision must be made on whether to invest in the product to make it more secure or replace it with a new product or service.

This type of report will also highlight the systems within the organisation that carry the greatest risk.

This is not a comprehensive list of vulnerability metrics, although it provides a reliable measure of technical debt. The tools your organisation uses to document security defects must have the necessary features to capture the relevant data and provide hooks that make the data available to other systems. Automated testing tools should provide the ability to export details of vulnerabilities into the organisation's vulnerability management system in real time. Dashboards should be placed around the engineering team area that display visual cues when a build fails due to a critical vulnerability being identified. Ultimately, everyone in the DevOps team must be encouraged to reduce technical debt.

Conclusion

In this chapter, I have described the final layer of Dev-SecOps, which incorporates security test automation into the value stream. There are many tools available to integrate this functionality into the delivery lifecycle, each with its own strengths and weaknesses. Using the security test pyramid as a guide, your engineers have the ability to choose the tools that are more suited to the technologies and frameworks they use, have the greatest accuracy and provide the best outcomes. Unfortunately, there is no silver bullet, so you are advised not to choose a single tool for your automated security solution. I prefer a combination of several tools, such as SAST, DAST or IAST, and SCA for applications security testing, but the decision ultimately resides within your DevOps engineering community with the support of the cybersecurity teams. Some security testing technologies (such as AST and SCA) are more mature than others (such as IaC and container image testing). Your security policy should reflect the possibility of a changing landscape to provide the best automation tools for your organisation.

In this chapter, I also highlighted the need for manual security testing within DevOps, either through scheduled pen testing or via bug bounty programmes. The key to their success is the integration with your DevOps practices to provide fast feedback loops and continuous improvement that ultimately turns manual security testing into validation that your automated processes are effective.

I also showed how to apply monitoring and alerting tools to your automated test strategy. These technologies can detect and report on unusual behaviour indicative of an attempt by hackers to exploit a vulnerability in your production systems. This feedback loop can be integrated into your value stream to provide greater vigilance against security defects.

Finally, I described how security vulnerabilities are managed to foster a culture of reducing them rather than one of recording them. Together with education and good design principles, this third layer completes the core foundation of DevSecOps.

SEVEN

Laying The Foundation

N ow that I have introduced you to the three layers of DevSecOps, the next step is to build the security foundations of DevSecOps. In this chapter, I explain how to increase the maturity of your DevOps practices to transition your organisation into a fully functional DevSecOps culture. I start with the important task of reducing your technical debt before moving on to how to implement the three layers, starting with education, then secure by design, and finishing with security test automation. This chapter describes how to measure the success of DevSecOps within your organisation.

The three layers of DevSecOps: laying the foundation for secure

Increase DevSecOps maturity

Depending on where your organisation is on its journey towards DevSecOps, the target is to reach a level of maturity in which everyone involved in their value streams understands their respective roles in the DevSecOps model. Development engineers, operations engineers, cybersecurity SMEs, product owners, architects and customers each have a part to play in delivering secure products that improve the focus, flow and joy of your organisation and delight your customers. The first step is reducing your technical debt.

Start reducing technical debt

Frankly: technical debt is crippling. At first, it may not appear to be a problem, but over time it starts to consume you and your organisation, impacting your ability to deliver new features as quickly as the market demands. The problem spirals out of control as you are forced to push a product out of the door as quickly as possible even though you know its list of defects (including security defects) has not been addressed. Indeed, you may not even have a list of defects, but you assume they must be there, having been given no time to test for them. In my experience, this is a common problem affecting many organisations. The intention is always to fix the bugs in the next release or deploy a patch to address the most critical vulnerabilities, yet the drive for new features and products outweighs the desire to reduce the number of defects, and the engineers sink into a spiral of releasing increasingly vulnerable software. The more defects that accumulate in the live system, the greater the security risk and the need to fix them. In the meantime, your engineers spend more time on working around existing issues, while giving up valuable feature development time to fix the more critical ones. The result is a buggy product that lacks the features to delight your customers and demotivates your engineering teams, whose work is constantly criticised.

This is technical debt, and it is a killer for any

organisation. Many companies, including big names such as Amazon and Microsoft, have had to stop their regular work to proactively address their technical debt. If they had not gone through this exercise, they would have drowned in a sea of vulnerabilities while watching their customers sail off in another direction, so your organisation must allocate time to prioritise and fix defects – particularly security defects – over new features. Only when the technical debt becomes manageable can the engineers start to work on features that will steer your organisation in the right direction.

Your organisation must identify the most common vulnerability types and train engineers to not only fix them but avoid introducing them in the first place. You also need to ensure your DevOps engineers review the design of applications and infrastructure to high-light areas of concern. A strategy to refactor the most vulnerable areas of the product should be drawn up, and it must include test- and behaviour-driven design principles to validate the implementation of the design. Security engineers and security champions should be directed to configure security testing tools so that they eliminate false positives and work with the DevOps engineers to show them how to resolve the vulner-abilities that have been reported. This exercise may take several weeks, during which time it is important to resist the temptation to start working on new fea-tures unless they are critical. By the end of the process to reduce technical debt to a manageable level, the DevSecOps *three layers* will form the foundation of its ongoing management while engineers focus on the business value of the products they work on.

Introduce an education programme

Security needs to be embedded in the culture of a successful company, which is why security education is the foundation layer that supports the DevOps organisational structure. Without systemic knowledge of security, it is more difficult to adopt good design principles and automation tools. Engineers will not be able to relate healthy design with security of the application and infrastructure they are working on, and the automation tools will be ineffective if engineers are unable to interpret their outputs. Educating your staff on security does not mean sending them off to complete a course on security; it needs to be aligned to a strategic policy.

Though there are many resources available for improving the security knowledge of the employees within your organisation, before buying online education licences, booking courses or sending staff to boot camps, you need to have a clear understanding of the outcome you want to achieve and how you measure it. As we have said, DevSecOps is not a team of developers, operations and security; DevSecOps is DevOps working securely. Each member of a DevOps team is responsible for security: for writing secure code, for creating secure infrastructure, and for continuously validating security. Therefore, you need to introduce a programme of education to teach employees to think about security first.

To introduce a programme of security education, identify which learning resources work best for each member of the DevOps organisation. You need to establish a process for recording the level of knowledge and skills within the teams; this metric must be maintained constantly to prevent entropy, which I cover later in this section. The teams should be encouraged to share their knowledge with their colleagues through activities such as peer programming, lunch-bag sessions and community events within the organisation. Team members must have access to collaboration tools to allow them to seek information from or share information with their colleagues.

There is a danger that organisations that enforce a particular style of education will limit the effectiveness of the programme. Traditional training methodologies practised in schools, where a teacher teaches and pupils listen, are institutionalised, but their effectiveness is questionable. This format is designed to meet a curriculum to give the majority of students the best opportunity to pass an exam. Yet, many children struggle with this and perhaps go on to achieve success when given the opportunity to learn new skills using other methods such as through an apprenticeship or vocational college course. The same applies in adult education: many people struggle with certain types of education. Therefore, you must adopt practices that allow your team members to learn security in a way that suits them and budget for this accordingly. Education should never be a chore; it needs to be fun for the individual.

As new people join the DevOps teams it is important to have processes in place to assess their level of knowledge of security so that an education plan can be drafted for them. It is common practice in larger organisations to ask new employees to complete several internal courses within the first few days of starting in a new position. Often, a security module is part of this eclectic mix of online learning, which once completed, provides assurances to the organisation that the employee is 'security trained'. In other words, new engineers are asked to take a twenty-minute course on the OWASP Top Ten, answer a few questions and that's the only security training they receive before being let loose on your product source code. Even if your organisation adopts a form of mandatory security training in order to satisfy auditors, it should not underpin a security education programme. Ideally, new starters should go through an assessment of their security skills to identify their strengths as well as their weaknesses. Recruiters should also determine their preferred method of learning. Using this information, you can construct a plan that allows them to work on their weaknesses using whichever process works best for the individual as well as share their strengths to educate outer group members.

All existing team members should be regularly assessed in a programme of continuous learning to make sure that your organisation is providing the best outcomes for its employees to improve the security culture within the DevOps team. Ultimately, you need to avoid entropy, which I cover next.

Education to combat entropy

Entropy has been described as the measure of disorder of a system. While each member of a DevOps engineering team has a certain skill level, over time these skills erode. The longer engineers go without practising or reading about a specific topic, the more likely they are to forget the finer details. Engineers may also leave the product teams or the organisation, taking their skills and knowledge with them. I call the process of losing this knowledge *entropy*. Prolonged entropy leads to gaps appearing in your knowledge; to overcome it, you should look to quickly establish some key processes to turn the tide of entropy within your organisation.

As we discussed in Chapter 4, a number of educational methodologies are available. The two that will help stem entropy are regular peer reviews and regular threat modelling. These processes can be established fairly quickly due to their low overheads, and they both provide forums to openly discuss potential design flaws that could highlight security risks within the applications and services being reviewed. The longer-term strategy involves setting up an education programme and designing a job retention scheme based on the continuous improvement of daily work and psychological safety.

If there are gaps in the collective knowledge of your DevOps engineering teams, target these weaknesses with activities that reintroduce the missing skills and knowledge. This could take the form of a hackathon, where teams focus on learning how to fix the most common categories of issues as a group. By the end of the event, the level of understanding among the team will be higher and your products more secure. Not only are these events easy to set up, they are often enjoyable for those involved and lead to positive outcomes.

Implement security design principles

The adoption of security design principles builds upon the education strategy, which means that individuals within each DevOps value stream must have a foundational understanding of the principles required to deliver secure products. There is no silver bullet for implementing good design principles – they evolve from a solid education strategy and commitment from the organisation – yet there are some strategies you can implement from the start. First of all, avoid putting all your golden eggs in one basket; in other words, you should not build one team of your best engineers while other teams include only less-knowledgeable individuals. Instead, you should build multi-skilled teams with staff of varying degrees of knowledge and ability. This allows the more experienced (and, therefore, often the most knowledgeable) DevOps engineers to work with the least experienced team members. This way, each value stream has the power to work as a unit to apply good design principles through peer reviews and knowledge sharing. Secondly, you should bring penetration testers into the DevOps teams to help identify potential design weaknesses during the early stages of feature development. Finally, you should introduce threat modelling exercises into the development lifecycle by inviting security experts to some of the daily scrum meetings. These three actions can be applied quickly and have an immediate impact on the quality of your DevOps engineers' products.

Implement security test automation

One of the goals of a successful DevOps organisation is to automate as much as possible. In Chapter 6, I described the various automation tools that your organisation can integrate into the value stream, but where does a DevOps organisation start? Some companies I have worked with over the years have bought licences of a commercial SAST tool, integrated it into the CI/CD pipeline and proclaimed they have a fully integrated automated security testing solution. Unsurprisingly, the number of security incidents they had to deal with did not decrease. In some cases, the situation worsened. Choosing which test automation solutions to adopt is driven by the design of your products. Container-based solutions have different testing requirements to non-container solutions. The languages and frameworks engineers use are also factors in the decision process. Ultimately, you need to build a test automation strategy.

Your strategy for implementing security test automation involves identifying and optimising the tools to offer the best outcomes, integrating them into your CI/CD pipeline and creating fast feedback loops. SAST is an easy technology to integrate but, as discussed in Chapter 6, care must be taken in reducing the number of false positives. Vendors of AST tools usually offer programmes to help your teams integrate their products into your continuous integration platform. These include professional services, training, installation support and helplines, which can help your DevOps engineers integrate the tools correctly and tune them to reduce false positives and increase accuracy of results. These programmes come at a cost, and each vendor is keen to exalt the value of their product over any others, so take care in choosing a vendor you feel comfortable working with. Many vendors offer free or low-cost products you can use to evaluate their offerings. Judgement of their products should be based on the evaluation made by your DevOps engineers. Once your teams have chosen a tool, they must commit to using it to analyse the applications for vulnerabilities. You must avoid apathy, particularly if the tool is too complex or is overly sensitive to false positives.

Application security testing tools focus on code that your engineers have written. But, as we saw in Chapter 6, this is a small proportion of the code in your organisation's software; the remainder consists of open source components written by people who do not work in your organisation. Many of these components integrate other open source components, which themselves integrate more open source components. Within this chain of dependencies, there are likely to be significant vulnerabilities that can put your organisation at risk if ignored. Therefore, I strongly suggest that, in parallel with integrating AST tools, you should implement dependency checking tools as your organisation's top priority to mitigate this risk. At minimum, your DevOps teams should perform software composition analysis of the applications they are developing and implement tools that scan images for potential vulnerable dependencies.

DevOps engineers should be able to experiment with SCA and AST tools to find the ones that work most effectively in your organisation; meaning the tools complement each other, have a high degree of accuracy, report low false positives (or can be configured to suppress false positives) and contain comprehensive details to fix the issues.

Experimentation

Test automation sounds simple to implement, but it is not as easy as you may think. I have seen many instances where security teams have acquired a test automation tool and thrown it to the operations engineers to integrate into the pipeline, while developers are told to start scanning their code for vulnerabilities. This approach is common, but ultimately it does not work. Organisations must adopt security test automation in a considered way. The first step is to invest time in understanding what each tool does and whether it is appropriate for your organisation. For example, does it support the languages and development frameworks the application engineers use? Stuart Gunter says, 'Organisations do not invest enough in assessing testing tools. They need to understand how they work by using them manually first in order to determine the best configuration, before applying it to the automation pipeline.' Operations engineers must resist the temptation to immediately integrate the tools into the pipeline without understanding their value to the security testing process first.

Using lean methodologies to experiment with different product features is a key benefit of DevOps. Teams can test hypotheses about their customers and products by developing experimental features that can be quickly rolled out to a small customer base in order to learn how customers interact with the new feature. If the feature is considered a success, it is refined and rolled out to the other customers; if it is a failure, it is simply removed from the product's offerings. This approach can also be applied to choosing the correct security test automation tools for your organisation.

Thus, your engineers should run small experiments with the various security test automation products to see which ones will offer the greatest benefits to your organisation. To start with, engineers should run manual scans using these tools and address the outputs of the scans within their regular daily activities. If running the manual scans gives your DevOps engineers the level of confidence that the tool is suitable for your organisation, you can make the investment to integrate the tools within the CI/CD pipeline.

Implement container image scanning

If your DevOps teams are using containers, it is imperative that container images undergo a high level of scrutiny to minimise vulnerabilities. In her book *Container Security*, Liz Rice explains that there are three types of image scanning within the CI/CD pipeline, and you should consider the best approach for your teams to integrate them. The first type of image scanning is carried out by developers within their IDE. This allows developers to fix issues before they push their images to the source code repository. Secondly, images are scanned during the build process, allowing operations engineers to validate images before they are deployed to the registry. If the images do not meet the required level of security, such as the discovery of a critical vulnerability, the build should at least provide information about the vulnerability that caused the failure. During the early stage of adoption, you may not want to fail the build; however, as your processes mature, you should adopt a policy to fail a broken build. Finally, once images have been built successfully and pushed to the registry, they should be scanned regularly, as often as daily, in case new vulnerabilities are discovered in a package used by the image.

Your DevOps engineers should work closely with cybersecurity experts to establish a policy for image scanning. As I mentioned in the 'What are containers?' subsection in Chapter 5 I strongly recommend that you discourage engineers from making changes to containers at runtime; you should make your images immutable. Even if vulnerabilities are discovered by scanning running containers in production, your policy should enforce rules to edit the image file and redeploy the image to the image registry. With a robust and efficient CI/CD pipeline, you can replace containers hosting vulnerabilities with clean containers as part of your container management policy configured within your chosen container orchestration tool.

Introduce policies carefully when working with containers. I have been in situations where a hard policy to fail vulnerabilities in the central image registry caused havoc when new vulnerabilities were discovered in images of the running containers. As these containers reached their production shelf-life, the orchestration tools tried to replace expired containers with new ones as mandated by the policy. However, these policies also blocked the promotion of the vulnerable images to production. As a result of these conflicting policies, containers were killed and not replaced, and applications started to fail in production, affecting the customers' experience.

Feedback loops

No matter which tools your engineering teams use to automate security testing, they should have efficient feedback loops to allow developers and operations to assess and fix vulnerabilities as quickly as possible. The fastest feedback loop for engineers writing code is within the IDE, where the issue can be fixed and re-scanned immediately before committing the code to the group code repository. If the scan passes or the engineer fails to scan the code at this stage, the CI/CD pipeline needs to perform its own scan after code has been committed and fail the build if the scans report any vulnerabilities. The reporting of failed builds provides better results if it is integrated within the engineering development tools, such as the IDE and the CI/CD pipeline, because in this scenario a breaking build due to a failed security scan notifies engineers that the build has failed. At this point, no-one should be able to check in any code until the build has been fixed.

Although there is a tendency to point the finger at the engineer who committed the build-breaking changes, it is important to realign the priority of the whole team to fix the build. Indeed, you should remove any remnants of blame culture from your organisation by encouraging positive outcomes from feedback loops.

Embolden your DevOps engineers to speak up without fear to improve the products they are developing. In the Toyota Production System, on which many principles of DevOps are based, the *Andon* cord system allows production line engineers to pull a cord to alert their peers when they see a problem. If the problem is not fixed immediately, the whole production line is stopped and the priority of every engineer is to fix the problem and restart the production line. Rather than being discouraged from pulling the cord, the engineers regard it as a key requirement to maintain the high quality of the components used in the production of Toyota's vehicles. A failed build needs to work in a similar way to the *Andon* cord. If the build fails on the CI/CD pipeline and the engineer is unable to fix the issue, the engineering process should halt until it can be fixed. The engineering team can decide whether to fix the vulnerability, re-configure the scan to ignore the issue (if it is a false positive), or roll back the change that broke the build, allowing more time to fix the issue. Re-configuring the scan is an important option: not all reported vulnerabilities are true positives, so engineers must be able to flag false positives to avoid further build failures resulting from the same vulnerability on the same component.

The feedback loop must also extend beyond the CI/CD pipeline in cases where manual penetration testing or bug bounty programmes identify vulnerabilities or when security defects are raised during an incident in production. Feeding this information into the defect

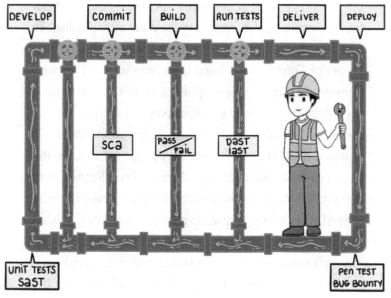

CI/CD pipeline: continuous feedback and test automation

management tool used by engineers allows progress to be prioritised and tracked. Avoiding the temptation to create an emergency response for each security incident is important since most incidents need to be investigated to establish the cause and effect before they can be prioritised. Nevertheless, if an incident suggests that an adversary is actively targeting your application, steps should be put in place to halt the attack without the need for software engineers to update significant parts of the source code or make changes to a live system. I was involved in dealing with a credential stuffing attack, where multiple attempts to log on to a system using known credentials obtained from breaches of other companies was halted by the simple introduction of a Captcha feature. This simple solution prevented automated 'bots' from attempting username and passwords combinations in a brute-force attack. In the meantime, work was scheduled to make the login journey more secure by introducing multi-factor authentication. The quick fix gave engineers time to redesign, develop and deploy a permanent solution in a controlled way.

Feedback from manual testers and bug bounty hunters tends to be in report format, and these static documents can be interpreted out of context. For example, a critical issue discovered by a penetration tester may not factor in other controls that mitigate the risk of the vulnerability, making it less likely to be discovered or have a lower impact on the business than indicated. Conversely, a low-rated issue may be treated as a low priority because the issue rating is taken at face value, but the impact on the business if the issue is exploited could be significantly higher. In both scenarios, the rating of the issue must be contextualised, documented and prioritised accordingly.

Measure and adjust

Implementing new processes should never be carried out blindly. Each step must be measured and adjusted to meet its original objective. The challenge in measuring a process is deciding which statistics to collect and at what frequency. Avoid collecting statistics that do not add value (often referred to as vanity metrics); they are produced to enhance the look of your adopted process, but they do not provide meaningful insight into the process. Any data collected should give engineering teams the ability to make quick decisions regarding what works well, to reduce time wasted on ineffective methods. I have often seen examples of vanity metrics used to make the wrong decisions. In DevOps security, I believe too much value is given to the number of vulnerabilities reported through security testing; the real value comes from knowing the type of vulnerabilities being reported, the impact they have on your customers and the time it takes to resolve them.

Obtaining metrics must give clues to the direction you need to take in making your products more secure, so be prepared to validate the measurements you are recording and adapt accordingly. Each metric must relate back to the customer value – measuring the number of vulnerabilities in a product adds no value to a customer, but measuring whether customers are affected by a specific type of attack does. If a new threat type is identified within your system or product, you need to adjust the data you collect so that its business value can be recorded.

This section offers some guidance on which data in each of the three layers provides the greatest value and how to collate and act on this data.

Measure knowledge and skills

Qualitative data is difficult to measure, and this is certainly the case for understanding the level of knowledge and skill within your engineering team. Traditionally, managers maintain a skills matrix that indicates each engineer's capability scores against a particular skill. Often the score is recorded in years or in a scale based on each developer's subjective self-assessment percentage rating. These matrices are usually unreliable and rarely updated, and I would not recommend scoring your team's skill level this way. You need to assess the skills as objectively as possible using quantifiable data, and there is greater benefit in assessing groups or teams rather than individuals. Collective knowledge is more important than individual knowledge, so you should explore options to capture data that implicitly measures the skills for whole teams. Data showing that a particular team is prone to writing code that contains a particular type of security vulnerability indicates that the team's understanding of that issue is worth exploring further. The assumption being made is that the code is vulnerable due to a weakness in the collective knowledge of the engineers in the team. To avoid jumping to the wrong conclusions, it is worth recording how often the team makes the same coding error rather than assuming that the first instance of the vulnerability is the result of a lack of knowledge or skill. Mistakes can be made, so it is important to identify consistent weaknesses over time rather than just a one-off anomaly.

Creating regular tournaments using training gamification software as discussed in Chapter 4 can also offer insight into engineers' capability. Running competitions regularly can identify potential weaknesses within a team and provide information for focused learning as part of the continuous improvement process.

Visualising the weaknesses within your DevOps capability is important and provides valuable information to help establish the goals of an education programme. You can visualise weaknesses by creating radar charts where each spoke represents a vulnerability type (perhaps based on the OWASP Top Ten lists), and the magnitude of each is determined by the frequency with which they appear in testing. When you plot these charts, you start to see where you may have strengths and weaknesses.

Automated testing gives engineers a fast feedback loop for discovering and fixing security defects. Each time automated security testing fails a build, it logs the cause of the build failure. By analysing this data over time, you can identify trends to determine which types of defects engineers are making most regularly, allowing you to choose which vulnerabilities to focus on during training. I will discuss measuring automated testing later in this chapter.

Be wary of using learning metrics during individual performance reviews. The motivation for collecting this data is to find gaps in the collective knowledge and advance the group skills that lead to the best outcomes for customers. When collective knowledge is directly linked to positive customer outcomes, individuals within the teams will be motivated to acquire the skills they need to provide the best outcomes. You can thus create a culture of learning by empowering teams to educate themselves.

Measure code quality

According to the *Consortium for Information & Software Quality* (CISQ), code quality is measured through the analysis of three distinct levels of the software. The first level is the *unit*, such as a function or method, which can be validated using code analysers embedded in the IDEs. The second level of analysis is performed against each of the individual *technology* stacks within the application. There are several open source and commercial tools that can perform this type of analysis to provide statistics on the hygiene of software for many common programming languages. These tools are often integrated into the CI/CD pipeline as each component is checked in and built. Some tools offer IDE plug-ins, allowing engineers to carry out analysis on their code prior to checking in their code. The third layer focuses on analysing the quality of code at the *system level*, in which the various components of the application are integrated into an end-to-end business application. This type of analysis is only performed within the CI/CD pipeline when the relevant components have been built and integrated. While the unit level measures the quality of coding practices, the technology and system levels measure the quality of architectural practices.

Each level is assessed based on four characteristics: reliability, performance, security and maintainability. It is important to measure all four characteristics and not just focus on security. As we have seen in Chapter 5, writing better-quality code implicitly improves the security of the application.

It is easier to measure code quality at the unit level using the types of tools described in Chapter 6. However, according to industry expert Richard Soley, CISQ warns us that although approximately 90% of coding errors are found within the unit level, they only account for 10% of defects found in production. On the other hand, technology and system defects account for 10% of coding errors, yet they cause 90% of production issues and take the longest to fix. Therefore, it is essential you measure code quality at the technology and system level to maintain good code hygiene and mitigate against the greatest risks to the organisation. Capturing these issues earlier in the development lifecycle is cheaper because the development team can fix them more quickly.

Measure test automation

The use of automated security testing tools should be embedded within the delivery lifecycle, as discussed in Chapter 6. You should continue to monitor the value of these tools by continually assessing their cost of ownership and effectiveness in addressing vulnerabilities in production. I've come across organisations that empha-sise rolling out tools to as many teams as possible and measuring product coverage as a yardstick for success. The problem with this approach is it gives you little insight into how the engineering teams use these tools or how effective they are in making your applications and services more secure. These automated tools often report a high number of vulnerabilities when they are first implemented. Until your teams have assessed and resolved each one, they are reporting false data, and this will skew your assessment of the tool's success. The value of test automation can be captured using different metrics, not only in terms of the accuracy of the automation tools but also in terms of its effect on the value stream.

Measure false positives

It is important to eliminate false positives from the total number of vulnerabilities that are discovered during testing. The false discovery rate is an indicator of the number of false positives versus the number of true positive findings:

```
false discovery rate (fdr) = number of
false positives (fp)/(number of true
positives (tp) + number of false positives
(fp))
```

This is a more useful indicator than looking exclusively at the total number of false positives recorded. The aim is to configure the tools to minimise the number of false positives in the result set without reducing the number of true positives. Monitoring the false discovery rate will give early indications on whether the number of true positives is affected by overly aggressive configuration of the tools. The target is to reduce the false discovery rate to 0 without reducing the number of true positives.

Estimate false negatives

Another important value is the number of false negative results *not* reported by the automated testing tools. These indicate that the testing tool has failed to recognise a genuine vulnerability within the code. This metric is more difficult to determine as it is inherently an unknown value; the tool cannot report on what it cannot find, so a heuristic approach is required. Because automated testing tools have different strengths, it is possible to indirectly observe the number of false negatives (and false positives) each tool records by using more than one product within the delivery pipeline. Tool A will identify genuine vulnerabilities that Tool B misses, and vice versa.

Once you ascertain the false negative value for each tool for a specific codebase, you can measure the sensitivity of the various automated testing tools. Calculating the sensitivity is based on the correlation between the number of true positives and false negatives. The more vulnerabilities the tool misses, the less sensitive the tool is. This is represented by the following formula:

```
sensitivity = number of true positives
(tp)/(number of true positives (tp) +
number of false negatives (fn))
```

Over time, engineers need to configure their testing tools to reduce the number of false negative results without negatively affecting the number of true positive results. The closer to 1 the sensitivity value, the more accurate the tool is in capturing vulnerabilities within the codebase.

Measure the value of application security testing

If the false discovery rate and sensitivity of application security testing are not measured and acted on, there is a danger that automated security scans will be perceived as adding no value. If they can be factored into the controls used for gated check-ins, limits on each value can be set to reduce the risk of running ineffective scans.

Not only is it important to know how effective the automated testing tools are at identifying genuine issues, it is also essential to know how they affect the value stream. If running security scans constantly leads to broken builds that require time to fix, engineers are likely to stop using the tools. This is particularly the case when the false discovery rate approaches 1 and the tools are seen as adding zero value; they fail to identify true vulnerabilities but flag a large number of false positives.

Running automated security scans can add time and cost to the CI/CD pipeline. To keep control of these values, various metrics can be used. Measuring the time it takes to run each scan within the pipeline allows engineers to fine-tune the scans accordingly. If the scans are taking too long to run, they may be running against a larger codebase than necessary. This is common when making incremental changes to the underlying codebase, but each time an engineer checks in code the scan runs against all the code rather than just assessing the deltas resulting from the code changes the engineer has made.

Time is also a factor in analysing the outputs of each scan as well as configuring the tools to reduce the number of false results. Treating these as work items that are managed through your task management tool allows time spent on these activities to be tracked. Any work that takes longer than necessary is a potential bottleneck, so monitoring these tasks lets you improve the efficiency of your engineering teams without reducing the quality of the security scans.

Hardware resources have limits on their capabilities: CPU and memory, bandwidth, and network traffic can reach maximum capacity, affecting the time it takes to run scans. Cloud solutions can help solve some of these problems by introducing a level of scalability; however, this comes at a financial cost, which ultimately affects the customer (either through pricing or quality of service). Maintaining an optimal level of throughput to avoid unnecessary costs and improve security outcomes therefore requires careful monitoring. Knowing how long scans take to run, how much CPU and memory they use and how frequently they run provides insights into the total cost of running security scans as well as providing data to allow engineers to optimise their use.

A measure of how much code is covered by security tooling is a useful metric to monitor as there could be vulnerabilities lurking in the code untouched by a scan. However, what if that code is redundant, is never executed or is part of a journey that your customers never use? Is it worth scanning the code? Yes, it is. Hackers find backdoors into systems through redundant code, so measuring code coverage should be factored into AST.

Measure the value of software composition analysis

Assessing the number of vulnerabilities within your code dependencies provides a measure of how well teams are limiting the number of weaknesses introduced through third-party components and libraries. This is important because many attacks against systems exploit known vulnerabilities. Dependency validation tools have their own strengths and weaknesses: some record only CVEs that are available within the US National

Vulnerability Database (NVD), which means they do not identify vulnerabilities in components that vendor-specific databases maintain, while some SCA tools report a vulnerability for an entire library of components despite being present in only one component. These anomalies can skew the metrics obtained from these tools. There is no easy solution to this problem; however, scanning source code and containers using multiple tools can provide more comprehensive datasets with which to work. Some vendors claim their tools are more accurate than others, but these proclamations are out of context with your implementation, so caution is needed.

Measure impact on the value stream

In addition to measuring how security test automation tools perform, a measure of how the tools impact the value stream is also valuable information. There are several ways these tools can affect the overall flow of the delivery pipeline; for example, they may require a lot of investment in time and resources that prevent engineers from working on features that add value to your organisation's applications and services. Therefore, it is essential to evaluate the costs of:

- Setting up and maintaining the tools within the CI/CD pipeline
- Identifying false positives and false negatives
- Assessing and triaging outputs for resolution
- Ongoing support (upgrades and licence renewals)
- Educating engineers on using the tools

These metrics provide an ongoing assessment of the value the tools bring to delivery teams, which can be factored into the impact analysis when considering the risks of using specific tools. Although it is important to identify vulnerabilities within your organisation's products and protect your customers' data by integrating security controls, there are valid reasons to choose alternative measures. For example, if the cost of integrating security tools is higher than the impact of a vulnerability within a product or service being exploited, the decision to implement less expensive tools is reasonable. In all cases, this decision must be based on evidence from the metrics available to you and not just on assumptions.

DevSecOps starts with people

The layers I describe in this book are ineffective without people. DevSecOps is a culture, it is a way of working and it involves bringing out the best in your people to deliver excellent value to – and instil confidence in – your customers. Security starts with the people of your organisation: without their commitment, you will not be as strong as you can be. DevSecOps builds on DevOps by laying down layers that strengthen the security within your DevOps practices. By bringing security skills into your value streams, you promote the ideal of locality; by developing security into your applications, your engineers experience the focus, flow and joy of developing high-quality products. Continuous education meets the ideal of continuous improvement and minimising knowledge entropy. Using tools to drive efficiency in resolving security defects instils psychological safety as engineers can freely discuss vulnerable code in production and find ways to fix them; and they allow your organisation to focus on the customer, whether internal or external.

Conclusion

In this chapter, I have described the steps to build the layers of DevSecOps. I started by describing technical debt and explaining that this can be a hindrance to continuous improvement as all your efforts are focused on workarounds and spiralling into perpetual inefficiency. By reducing your technical debt and tackling open vulnerabilities, poorly designed architecture and code, you can start to mature your DevSecOps practices with confidence.

We then explored how to integrate an education programme, how to instil the need for designing for security, and the efficient adoption of test automation practices, followed by a guide on how to measure the three layers within your organisation so that you gain the most from them. Finally, I briefly tied all the layers back to the five ideals by explaining that security in DevOps starts with your people.

EIGHT

Summary

This book has introduced you to an approach for adopting DevSecOps within your organisation. In it, I presented three layers that form the foundation of DevSecOps. No matter how far down the DevOps journey you are, you can apply the principles I outlined. Each layer builds on the previous one to form a solid foundation of security principles that support the DevOps philosophy. I started this journey talking about the bedrock of your DevSecOps platform, which is security education. The collective knowledge of your DevOps teams must incorporate the skills required to build secure applications and their supporting infrastructure. Without the necessary understanding of the threats their systems face, your DevOps engineers are already facing a losing battle against the malicious actors who want to steal your organisation's data and disrupt services for their own gain.

There are numerous learning methodologies that

offer different benefits based on personal preferences, existing company culture and budget. The most effective form of education is one in which individuals are able to adapt their learning according to the work they are doing and practise their new skills, while the worst methods are those in which employees are sent off to complete a course on a topic that may not be perceived as relevant to their day job. Your organisation should look to establish a culture of learning, where individuals are encouraged to openly share ideas, help their peers and learn from incidents without any fear of repercussions.

Consider filling empty walls with posters that present easy-to-read details of the most common security weaknesses, buy books and subscribe to online resources, giving your DevOps engineers free access to them at any time. Give your teams the freedom to experiment and learn from their experiences; give them space and time to continue their security education by letting them have the chance to attend events that interest them and supporting local forums where your staff can share ideas with their peers. Make learning fun by setting up tournaments or gamifying their workloads.

Ultimately, knowledge is power, which translates into a working environment where key decisions are based on the collective knowledge of your teams and not on the seniority of the decision maker. By establishing a culture of security education, you reduce the risk of entropy by preventing the slow haemorrhaging of security knowledge and skills from your organisation.

Once you have developed a sustainable security education programme, the next level is to apply that knowledge to develop applications and infrastructure in which security is at the heart of every decision made by everyone involved in the value stream. Your DevOps teams should have the freedom to design, develop and deploy security features that protect the interests of your customers and your employees. This means they should always endeavour to write clean code that is easy to maintain, less likely to contain defects and is built within a robust architecture. The infrastructure that supports the development of your applications, as well as the hosting of your products, must have security at its core, from access controls based on least privilege to controls that protect the data as it passes through the ecosystem. Security is best served by simplicity, so make the design of the applications and services less complex. You should promote the practice of incremental changes and small work packages that are easier to design and code, rather than demand overly complex deliverables that are difficult to test and integrate.

Finally, you must integrate automated security testing into the DevOps workload. These tools must be fit for purpose and effective. Their goal is to assert the first two layers of DevSecOps by validating that your engineers know how to secure their applications and infrastructure and have adopted good security design principles throughout. There are many forms of security testing, so you should create an environment that empowers your engineers to choose the products that provide the best outcomes. There are many factors involved in this decision, from the technologies and frameworks your teams use to the accuracy and effectiveness of the testing tools. This is not a decision that should be made centrally by a security department but collaboratively between security subject matter experts and the DevOps engineers. Once you have adopted these tools, engineers should continue to evaluate their effectiveness at identifying security issues so that they can iteratively improve their performance.

Through security test automation within the continuous integration pipeline, DevOps engineers can deliver features at speed with minimal security risk. The feedback loops from automated testing allow engineers to identify and fix features before they occur in production. When combined with an effective education policy and good design principles, security test automation enhances continuous learning through these feedback loops and provides an environment where security is at the heart of software delivery without negatively impacting flow and customer outcomes.

Continuously monitoring your three DevSecOps layers gives you the opportunity to hone the skills your DevOps engineers need as well as improve the quality of your products. Ultimately, this adds value to your bottom line by giving your customers the confidence to return to your applications and services. With security education, security design principles and automated security testing in place, you minimise the risks of pushing defective code to production. Equally, if your engineers identify vulnerabilities in production, they have the capability to fix them cleanly and efficiently to reduce the impact on your customers.

You may be wondering where you start in laying down this DevSecOps foundation. First of all, you need to define what DevSecOps looks like in your organisation. You also need to decide how your teams begin to apply the layers to gain the most benefit. If you try to develop a comprehensive education programme before moving on to secure by design, the task may be too large and too disruptive to be effective; therefore, start small. Identify the value streams where you can introduce an education programme specific to their requirements and where you can integrate security as a design feature from the start. Then look at the tools these value streams should assess based on the technologies and frameworks they use. Over time, each layer will start to expand into more value streams until more of the organisation is working with DevSecOps principles.

As the adoption of DevSecOps practices widens and you realise the benefits, continue to measure what works and what does not, keeping the more effective processes and tools while discarding those that are less effective. Over time, your organisation will have laid a solid foundation for integrating security and DevOps into Dev...Sec...Ops.

References

Austin, E-M, interviewed August 2020

Austin, E-M, and Miller, M, 'The Ladies of London Hacker Society' [podcast], DevSecOps, 2019, www.devsecopsdays.com/devsecops-podcast-gallery/podcasts/the-ladies-of-london-hacker-society

Beck, K, *Extreme Programming Explained: Embrace change* (Addison-Wesley, 2000)

Beck, K, et al, 'Manifesto for Agile Software Development', 2001, https://agilemanifesto.org

Center for Internet Security, 'CIS Benchmarks', (nd), www.cisecurity.org/cis-benchmarks

Cohn, M, *Succeeding with Agile: Software development using Scrum* (Addison-Wesley, 2010)

Coulstock, Ray, interviewed August 2020

Das, Debasis, interviewed August 2020

Evans, E, *Domain Driven Design: Tackling complexity in the heart of software* (Addison-Wesley, 2004)

Fowler, M, 'Xunit' [article], 2006, https://martinfowler.com/bliki/Xunit.html

Gamma, E, et al, *Design Patterns: Elements of reusable object-oriented software* (Addison-Wesley, 1995)

Gunter, Stuart, interviewed August 2020

Have I Been Pwned, www.haveibeenpwned.com

Hern, A, and Pegg, D, 'Facebook fined for data breaches in Cambridge Analytica scandal', *Guardian*, 2018, www.theguardian.com/technology/2018/jul/11/facebook-fined-for-data-breaches-in-cambridge-analytica-scandal

Izrailevsky Y, and Tseitlin, A, 'The Netflix Simian Army' [blog post], Netflix Technology Blog, 2011, https://netflixtechblog.com/the-netflix-simian-army-16e57fbab116

Johnsson, D, et al, *Secure by Design* (Manning Publications, 2019)

Kim, G, *The Unicorn Project: A novel about developers, digital disruption, and thriving in the age of data* (IT Revolution Press, 2019)

Kim, G, et al, *The DevOps Handbook: How to create world-class agility, reliability, and security in technology organizations* (IT Revolution Press, 2016)

Madou, Matias, interviewed August 2020

Martin, RC, *Clean Code: A handbook of Agile software craftsmanship* (Prentice Hall, 2008)

MITRE, '2019 CWE Top 25 Most Dangerous Software Errors', CWE, (nd), https://cwe.mitre.org/top25/archive/2019/2019_cwe_top25.html

National Cyber Security Centre, 'CHECK – penetration testing', 2019, www.ncsc.gov.uk/information/check-penetration-testing

Nieles, M, et al, 'NIST Special Publication 800-12, Revision 1 – An Introduction to Information

Security', US Department of Commerce National Institute of Standards and Technology, 2017, https://nvlpubs.nist.gov/nistpubs/SpecialPublications/NIST.SP.800-12r1.pdf

OWASP Projects, (nd), https://owasp.org/projects

OWASP Security Knowledge Framework, (nd), https://owasp.org/www-project-security-knowledge-framework/

OWASP Top Ten website, https://owasp.org/www-project-top-ten

OWASP Top Ten 2017 PDF document, (nd), https://github.com/OWASP/Top10/raw/master/2017/OWASP%20Top%2010-2017%20(en).pdf

Özil, G, @girayozil 'Ask a programmer to review...' [Tweet], February 27 2013, https://twitter.com/girayozil/status/306836785739210752

Rice, L, *Container Security: Fundamental technology concepts that protect containerized applications* (O'Reilly Media, 2020)

SANS, 'Mistakes People Make that Lead to Security Breaches: The seven worst security mistakes senior executives make', 2005, www.sans.org/security-resources/mistakes

Shostack, A, *Threat Modeling: Designing for security* (Wiley, 2014)

Soley, R, 'How to Deliver Resilient, Secure, Efficient and Easily Changed IT Systems in Line with CISQ Recommendations', (nd), www.omg.org/news/whitepapers/CISQ_compliant_IT_Systemsv.4-3.pdf

Synopsys, 'Open Source Security and Risk Analysis

Report', 2020, www.synopsys.com/content/dam /synopsys/sig-assets/reports/2020-ossra-report.pdf

Tzu, S, *The Art of War* [Translated by John Minford] (Penguin Classics, 2008)

Willis, J, 'The Andon Cord', IT Revolution, 2015, https:// itrevolution.com/kata

Further Reading

A number of individuals, and their writings, have influenced my career and this book. The following list of resources helped me throughout my journey:

Beck, K, et al, 'Manifesto for Agile Software Development', 2001, https://agilemanifesto.org

Bismut, A, 'Dynamic Threat Analysis for Container Images: Uncovering hidden risks' [blog post], Aqua Blog, 2020, https://blog.aquasec.com/dynamic -container-analysis

Ellingwood, J, 'Understanding IP Addresses, Subnets, and CIDR Notation for Networking' [online tutorial], DigitalOcean, 2014, www.digitalocean.com /community/tutorials/understanding-ip-addresses -subnets-and-cidr-notation-for-networking

Forsgren, N, et al, *Accelerate: The science of lean software and DevOps* (IT Revolution Press, 2018)

Humble, J, and Farley, D, *Continuous Delivery: Reliable software releases through build, test, and deployment automation* (Addison-Wesley Professional, 2010)

Humble, J, et al, *Lean Enterprise: How high performance organizations innovate at scale* (O'Reilly Media, 2015)

Izrailevsky, Y, and Tseitlin, A, 'The Netflix Simian Army' [blog post], Netflix Technology Blog, 2011, https://netflixtechblog.com/the-netflix-simian -army-16e57fbab116

Johnsson, D, et al, *Secure by Design* (Manning Publications, 2019)

Kim, G, et al, *The Phoenix Project: A novel about IT, DevOps, and helping your business win* (IT Revolution Press, 2013)

Martin, KM, *Everyday Cryptography: Fundamental principles and applications* (Oxford University Press, 2012)

Martin, RC, *Clean Agile: Back to basics* (Pearson, 2020)

Martin, RC, *Clean Code: A handbook of Agile software craftsmanship* (Prentice Hall, 2008)

Matthee, M, 'SANS Institute White Paper – Increase the Value of Static Analysis by Enhancing its Rule Set', SANS Institute, 2020, www.sans.org/reading-room /whitepapers/securecode/increase-static-analysis -enhancing-rule-set-38260

National Cyber Security Centre, 'CHECK – penetration testing', 2019, www.ncsc.gov.uk/information/check -penetration-testing

Nieles, M, et al, 'NIST Special Publication 800-12, Revision 1 – An Introduction to Information Security', US Department of Commerce National Institute of Standards and Technology, 2017, https://nvlpubs .nist.gov/nistpubs/SpecialPublications/NIST.SP.800 -12r1.pdf

Rice, L, *Container Security: Fundamental technology concepts that protect containerized applications* (O'Reilly Media, 2020)

Rice, L, and Hausenblas, M, *Kubernetes Security: Operating Kubernetes clusters and applications safely* (O'Reilly Media, 2018)

Ries, E, *The Lean Startup: How constant innovation creates radically successful businesses* (Portfolio Penguin, 2011)

Schein, EH and PA, *The Corporate Culture Survival Guide* (Wiley, 2019)

Shostack, A, *Threat Modeling: Designing for security* (Wiley, 2014)

Soley, R, 'How to Deliver Resilient, Secure, Efficient and Easily Changed IT Systems in Line with CISQ Recommendations', (nd), www.omg.org/news/whitepapers/CISQ_compliant_IT_Systemsv.4-3.pdf

Tzu, S, *The Art of War* [Translated by John Minford] (Penguin Classics, 2008)

Vehent, J, *Securing DevOps-Safe Services in the Cloud* (Manning Publications, 2018)

Vocke, H, 'The Practical Test Pyramid', martinfowler.com, 2018, https://martinfowler.com/articles/practical-test-pyramid.html

Acknowledgements

I could not have written this book without the help of many individuals who inspired me, gave me insights and kept me motivated to the end. In particular, I am grateful to my mentor Rob Kerr, who showed me that writing a book was within my potential and gave me encouragement throughout the writing of the book. He also pointed me in the direction of Alison Jones, who helped me formulate the book and give it structure, for which I am truly thankful.

I am also indebted to Johnathan Haddock, whose feedback on the manuscript was invaluable to correct a number of inaccuracies that had crept into the original draft. I was also encouraged by regular feedback from Chris Wilmott and Seb Ben M'Barek, who gave different perspectives on some of the topics covered in the manuscript.

During the writing process, I received help from a number of experts who I consulted on technical matters covered in the book. My gratitude is extended to Eliza-May Austin, Steve Giguere, Jamel Harris, Stuart

Gunter, Matias Madou, Ray Coulstock, Debasis Das and Chris Rutter.

The drawings in this book were created by the talented Crystal Tamayo, injecting some visual humour into the prose and for that, I am very grateful.

Thank you to the talented team at Rethink Press, especially to Maya Berger whose edits made this book so much better, Joe Gregory for translating my vision into the wonderful cover design, and Kate Latham for managing the whole publication process and for keeping it on track.

I have been inspired by Michael Man, who has been a big supporter of my journey in DevSecOps over the many years I have known him. He has given me opportunities to share ideas presented in this book at his brainchild community forum, the award-winning DevSecOps London Gathering. Thank you, Michael.

Finally, I would like to thank my family for their support and, of course, my lovely partner Caz for encouraging me to reach for the sky!

The Author

 Glenn Wilson is the Chief Technology Officer and Founder of Dynaminet, a company that specialises in consulting organisations on DevSecOps and secure Agile practices. He is an experienced development and security consultant who has worked for over twenty years in the IT industry across multiple sectors, including the highly regulated financial markets, payment providers, insurance and healthcare. Glenn has driven digital transformation programmes to integrate secure Agile and DevOps practices into the software delivery lifecycle, and he has provided security consultancy to large international banks.

Glenn is an active member of several professional community groups in the UK and Europe and has given talks on security at a number of their organised events, including the award-winning DevSecOps London Gathering group and the British Computer Society. He is also active in local digital marketing groups and software development groups, in which he has given talks on a number of security topics to both technical and non-technical audiences. In recent years, he has published whitepapers on security testing and cloud security, becoming an authority on these subjects.

Glenn has the unique ability to speak confidently about DevSecOps in a language that is clearly understood by non-technical clients, while remaining conversant in the technical complexities of his domain.

You can reach Glenn at:

✉ glenn.wilson@dynaminet.co.uk
🌐 https://dynaminet.com
🔗 www.linkedin.com/in/glennwilson
🐦 @GlennDynaminet